Britannica®

Encyclopædia Britannica, Inc., is a leader in reference and education publishing whose products can be found in many media, from the Internet to mobile phones to books. A pioneer in electronic publishing since the early 1980s, Britannica launched the first encyclopedia on the Internet in 1994. It also continues to publish and revise its famed print set, first released in Edinburgh, Scotland, in 1768. Encyclopædia Britannica's contributors include many of the greatest writers and scholars in the world, and more than 110 Nobel Prize winners have written for Britannica. A professional editorial staff ensures that Britannica's content is clear, current, and correct. This book is principally based on content from the encyclopedia and its contributors.

Introducer

Cordelia Fine studied Experimental Psychology at Oxford University, followed by an M.Phil in Criminology at Cambridge University. She was awarded a Ph.D in Psychology from University College London. She is currently a Research Associate at the Centre for Applied Philosophy & Public Ethics (CAPPE) at the University of Melbourne, Australia. She is the author of the highly acclaimed *A Mind of Its Own: How Your Brain Distorts and Deceives* that was short-listed for the 2007 Royal Society Book Prize. Her website can be found at: www.cordeliafine.com.

Also available

The Britannica Guide to the 100 Most Influential Americans

The Britannica Guide to the Ideas that Made the
Modern World

The Britannica Guide to Modern China

THE **Britannica**® GUIDE TO

THE
BRAIN

A guided tour of the brain –
mind, memory, and intelligence

ROBINSON

RUNNING PRESS
PHILADELPHIA · LONDON

Constable & Robinson Ltd
3 The Lanchesters
162 Fulham Palace Road
London W6 9ER
www.constablerobinson.com

Encyclopædia Britannica, Inc.
www.britannica.com

First published in the UK by Robinson,
an imprint of Constable & Robinson, 2008

Text © 2008 Encyclopædia Britannica, Inc.
Introduction © 2008 Cordelia Fine

A copy of the British Library Cataloguing in Publication
Data is available from the British Library

UK ISBN 978-1-84529-803-6

3 5 7 9 10 8 6 4 2

First published in the United States in 2008 by Running Press Book Publishers
All rights reserved under the Pan-American and International Copyright Conventions

9 8 7 6 5 4 3
Digit on the right indicates the number of this printing

US Library of Congress Control Number 2007936633
US ISBN 978-0-7624-3369-8

Running Press Book Publishers
2300 Chestnut Street
Philadelphia, PA 19103-4371

www.runningpress.com

Printed and bound in the EU

CONTENTS

Part 3 What Happens When Things Go Wrong?

LIST OF ILLUSTRATIONS

INTRODUCTION

Questions from the Frontier of Knowledge
Cordelia Fine

Introducing a three hundred page guide to the brain feels a little like grandly opening the door to a magnificent mansion that one does not own. I can invite you to take a tour of the cerebral cortices, modestly enquire if you'd care to see the inner workings of a neuron, or wonder out loud whether you might be interested in browsing the large collection of brain imaging techniques. But having assumed a propriatorial air over all that has been learned about the brain and mind, and is laid out in this Britannica Guide, I am not even going to pay the compliment of glorying in these achievements of others.

Instead, I want to focus on some of the new questions and issues that now flourish in the shadows of the many bright successes of neuroscience. Because, for some, the crossing of another frontier of research – rather than providing a good excuse to crack open the champagne – instead just throws up yet another tricky problem to solve. Scientists, for instance, might think it rather fine to uncover the neurochemical basis of love. But as a result, ethicists now need to address the possibility of love being neurochemically enhanced – or even created.

Is there something objectionably inauthentic about refreshing wilted love with a splash of synthetic oxytocin? Or is a "love drug" simply the psychopharmaceutical equivalent of the weekend away – the perfect solution for busy couples who want to stay together, only he can't stand the way she nibbles toast, she shudders at every slurp that echoes around his teacup, and an oxytocin-depleting spat over the hotel breakfast table is probably all but guaranteed?

Similarly, progress in developing medication for disorders of attention and alertness brings with it great scope for the treatment of conditions that can make life horribly hard, such as ADHD and narcolepsy. But even before the applause for the neuroscientists' contribution to alleviating human suffering has died away, the possibility of those same drugs being used by people without any clinical problems, to enhance focus or wakefulness to "better than normal" levels, needs to be addressed.

Will we, ask some ethicists, soon need blood-testing for illicit performance-enhancing drugs in the examination room as well as on the sports field? Others worry that the development of drugs that extend wakefulness will present a temptation too hard to resist for the type of employer who regards an employee's need to lie unproductively unconscious for several hours a night as less of a fact of nature than an unacceptable drain on the bottom line.

Any medical advance, of course, raises difficult questions about fairness of access, risks, and the potential for abuse. But as philosopher Neil Levy points out in the inaugural issue of the journal *Neuroethics*, neuroscience also throws up challenges that go well beyond the normal bioethical issues raised by progress in medical research:

> Whereas medical advances, important as they are, deal with our *bodies*, neuroscientific discoveries promise – or threaten – to reveal the structure and functioning of our *minds* and, therefore, of our souls . . . And there really seems to be a sense in which neuroscience (and the other sciences of the mind) is stripping back the mysteries of the mind in sometimes disturbing ways, threatening our notion of ourselves as autonomous, rational and moral beings.

For example, neuroscience's promise – or threat – to show how behaviour arises from causal, physical events in the brain can unsettle our beliefs about free will and moral responsibility. When a sleep-deprived mother of three small children tips a bowl of soggy cereal over the head of her insensitive husband we may sympathize, yet nonetheless hold her responsible for her action. But what if, some day, we could see the effects on the mother's brain of being woken at 1 a.m., 2 a.m. and 5 a.m., combined with neurological changes brought about by the physiological craving for an early morning cup of tea, the realization that the last of the milk has just been used to make a luxuriously deep lake for her three-year-old's Weetabix-boat, followed immediately by the observation that (while she was recklessly absent from the kitchen in order to clean her teeth) her youngest has begun an ambitious finger-painting project with the strawberry jam? Suppose neuroscientists could also chart, in real time, the neural effects of a husband, fresh from a shower, who enquires whether she has forgotten that he takes his tea with milk and, by the way, wouldn't it be a good idea to get that jam off the curtains before it stains? As the scene unfolds, the neuroscientists look at each other edgily, knowing even before she does that the woman is about to reach for the cereal bowl.

This sort of scenario makes us nervous. "The underlying worry is that . . . neuroscience will lead us to see the 'universe within' as just part and parcel of the law-bound machine that is the universe without", is how philosopher Adina Roskies has described the "hand-wringing" stimulated by the extra-ordinary potential of neuroscience. And if neuroscience, like an omniscient God, can foresee how we will react in a situation, "how can we preserve our intuitions about freedom and its conceptual partner, moral responsibility?"

But, as Roskies and others note, fears that neuroscience poses a special threat to free will are misplaced. There is nothing new about the idea that human decision-making is purely mechanical and determined. People have long worried that in a deterministic world, everything is the product of the starting state of the universe and the laws of nature – including our mental states and the brain activity that causes them. All neuroscience does, by way of its dramatic mechanistic explanations of behaviour, is bring to life this abstract philosophical idea.

Neuroscience is therefore considered something of a red herring by those who think that responsibility calls for ration-ality and intent, rather than the existence of other possible ways of acting. What counts is whether you can work out what you should or shouldn't do, and then act on the conclusions of those deliberations. From this point of view, "all behaviour may be caused in a physical universe, but not all behaviour is excused, because causation per se has nothing to do with responsibility", as legal scholar Stephen Morse argues. Neu-roscience may terrify us, as it "relentlessly exposes the numer-ous causal variables that seem to toss us about like small boats in a raging sea", says Morse, but it offers no threat to what is most important about ourselves: "that we are conscious, intentional, and rational creatures".

But some argue that neuroscience may yet change our views. Cognitive neuroscientists Joshua Greene and Jonathan Cohen speculate that, as increasingly sophisticated neurological explanations penetrate our lives, neuroscience may eventually overthrow the intuitive sense of free will that runs deep in all of us. None of us, after all, has control over the many causes, "from the genes you inherited, to the pain in your lower back, to the advice your grandmother gave you when you were six", that act through our brains to make us "who [we] are and what [we] do". Neuroscientific research may serve to reinforce the idea that all of our beliefs and desires are the products of forces beyond our control, and do this so compellingly that we can no longer ignore it. We will grow up "completely used to the idea that every decision is a thoroughly mechanical process, the outcome of which is completely determined by the results of prior mechanical processes".

As a result, we may become uncomfortable with the law's stance that people can be held morally responsible for their actions. Perhaps we will demand a more compassionate perspective on criminal behaviour in which, to use a French proverb, "to know all is to forgive all" and deterrence, rather than retribution, is the primary purpose of punishment. Neuroscience, argue Greene and Cohen, is no problem for the law as it stands – but may eventually be the intuition-changing force that transforms it. "For the law," suggest Greene and Cohen, "neuroscience changes nothing and everything".

Recent progress in the cognitive sciences is potentially no less controversial. In particular, those who prefer a self-conception that includes a rational, reflective self that is firmly in charge may not like what is to be found there. A particular specialty of contemporary social psychology is the demonstration of "implicit cognition" or the "broad range of mental life now understood to proceed without the encumbrances of awareness, intention, and control", as psychologists Brian Nosek and Jeffrey Hansen have put it. It turns out that our judgements and behaviour can be affected, in ways we don't realize, when implicitly held and unconsciously acquired associations in the mind are "primed" by environmental stimuli. For instance, the implicit mind readily and effortlessly picks up the patterns of what society shows (or merely says) it means to be elderly. An elderly person on the bus then primes all those associated concepts, making stereotyped attributes of old people more likely to influence subsequent judgement and behaviour. It is for this reason, as research has shown, that unsuspecting volunteers walk more slowly down a corridor after solving anagrams about being elderly. Similarly, without even realizing, volunteers who have just examined a picture of a library they think they are going to visit, speak in more hushed tones than those who have just inspected an image of a railway station.

In a way, it is a marvel that the mind is so very attuned to subtle environmental cues. We may not even be consciously aware of whatever it is in the environment that is affecting our behaviour (as when psychologists use subliminal stimuli). But, for humankind unsparingly observed under the social psychologists' microscope, this research brings with it more than an embarrassing sense of having been duped. For this is now at the core of a debate over just how much of a role conscious intention actually plays in our behaviour. Unfortunately, there is a good case to be made that it is only insofar as we act in line with our uniquely human beliefs and values that we can justly think ourselves rational. Whenever researchers find evidence that our behaviour is controlled by the environment rather than conscious intention, they potentially offer support to the troubling idea that we are not the rational animals we thought we were.

Nowhere is this problem more apparent than in realms of life where our explicit values and implicit impulses tend to conflict and our conscious control capacities are, for some reason, temporarily at low ebb. In these situations, research suggests, it is our implicitly held associations that guide behaviour. Distracted, we may laugh at racist jokes our conscious selves find distasteful. Under time pressure, we may choose expensive, branded products even though we actually prefer cheaper generic goods. Mentally fatigued, we may eat more candy than we "really" want to. Unfortunately, these lapses are not just bad for our wallets or waistlines: they are not so good for our conception of ourselves as creatures guided by cool reason and logical deliberation. There is, after all, something disquietingly irrational about behaviour, triggered by implicit impulses, that we neither consciously intend nor endorse.

What is more, there are further difficult questions about moral responsibility to address when research suggests that we can remain blissfully unaware of the unwanted effects on behaviour stemming from involuntary and unconscious mental processes. Should we be blamed if we sincerely intend not to be racist, but don't realize how our judgement is tainted by implicitly held stereotypes? Such questions emphasize Brian Nosek's point that "understanding the psychology of individuals involves what people believe about themselves, and what happens in minds without explicit permission". Empirical research will continue to interact with long-standing philosophical issues, and the answers may force us to reconsider our ideas about responsibility, culpability, and even what constitutes what we think of as "me".

*   *   *

And finally, the increasing recognition of the responsiveness of both brain and mind to the experiences and culture in which it develops and functions underscores the fact that although the brain, neatly tucked away in its hard skull, is physically separated from the rest of the world, the two are nonetheless inextricably linked. As cultural psychologists Shinobu Kitayama and Dov Cohen note:

> the psyche is . . . not a discrete entity packed in the brain. Rather, it is a structure of psychological processes that are shaped by and thus closely attuned to the culture that surrounds them . . . the mind cannot be understood without reference to the sociocultural environment to which it is adapted and attuned.

As cognitive neuroscience continues to explore the neural mechanisms that underlie memory, perception, self-identity, judgement and action, it should never lose sight of the fact that every brain has been individually influenced by culture, experience and context. Exciting new evidence of the brain's "neuroplasticity" most obviously underlines this point, with its demonstration of structural and functional changes in the brain in response to experience. But perhaps, increasingly, exploration of ways in which culture and social context can change the brain and how it works will force recognition of just how inseparable are the brain and culture. Cognitive neuroscientists have, for example, observed how the way that a mental rotation task is presented to women (as either a threateningly "male" spatial reasoning task or a more feminine-sounding "perspective-taking" test) affects which brain regions are recruited – and even how well the women do at it. Similarly, functional imaging studies suggest that parts of the brain involved in challenging, attention-demanding cognitive tasks can be "fatigued" by taxing social interactions. The effects of Western versus Eastern cultural context have been seen in the brain scanner too, when culturally-based differences in preferred judgement tasks translate into striking differences in brain activations. It does not take too much imagination to realize that this evidence of the powerful influence of culture and situation on brain and mind is as potentially unsettling as it is liberating. Our thinking may be less independent than we realize.

The more we know about our minds and brains the less, it seems, we know ourselves. This is, however, a small price to pay. Fortunate indeed is the person whose life runs untouched by psychiatric, psychological or neurological problems. For the sake of everyone else, we can cheer on the neuroscientific and psychological advances that offer so much hope of treatment, even of prevention, and ride out the philosophical uncertainties. For, as Greene and Cohen point out, human nature almost always, in the end, trumps existential angst. You might, for example, decide that the balance of evidence and argument undermines the idea that you have any causal efficacy in the world. Nonetheless, you will carry on being you for the simple reason that:

> you are a human, and that is what humans do. Even if you decide, as part of a little intellectual exercise, that you are going to sit around and do nothing because you have concluded that you have no free will, you are eventually going to get up and make yourself a sandwich.

As you will see in this guide, we have learned a great deal about ourselves in recent years, with the promise of much more to come. We need to be ready to tackle this new self-understanding, and all that it brings. Scientific advances in neuroscience and the cognitive sciences will continue to give rise to difficult ethical dilemmas, to cast old questions in a new light, and to provoke questions that go to the very core of who we are and why we do what we do.

PART I

WHAT IS THE BRAIN?

I

THE BRAIN: A GUIDED TOUR

The human brain is a miraculous organ. It regulates thought, memory, judgement, personal identity, and other aspects of what is commonly called the mind. It is the seat of our hopes and dreams, and our imaginations. It is our centre of learning. It also regulates aspects of the body – including body temperature, blood pressure, and the activity of internal organs – to help the body respond to its environment and to remain healthy. In fact, the brain is considered so central to human well-being and survival that in many parts of the world the death of the brain is considered to be legally equal to the death of the person. The brain is said to be the most complex living structure known to the universe.

The brain and the spinal cord make up the central nervous system, which can be likened to a super computer – processing and communicating the information that controls all of the body's many functions. The brain is generally defined as the part of the central nervous system that is contained in the skull. The spinal cord extends from the base of the brain and is contained within the vertebral canal. The brain controls the activities of the body and receives information about the body's inner workings, and about the outside world, by sending and receiving signals via the spinal cord and the peripheral nervous system every second. The brain receives the oxygen and food it needs to function by way of a vast network of arteries that carries fresh blood to every part of the brain.

The brain of a human adult weighs about 3 pounds (1.4 kg). It looks rather like a mushroom contained within the skull. The cap of the mushroom – the very top of the brain – is the cerebrum, and the stem of the mushroom – the part of the brain attached to the spinal cord – is the brainstem. At the back of the head, lying between the brainstem and the cerebrum, is the cerebellum.

Generally, the lower a part of the brain is within the skull, the more primitive and basic its function and the less likely it is that conscious control is involved in regulating the function. Thus the brainstem, the lowest part of the brain, is involved with the most fundamental processes, such as relaying information between parts of the brain or between the brain and the body, and regulating basic body functions. The cerebellum, behind it, controls balance and coordination. The cerebrum, in the topmost region of the skull, is the "thinking" part of the brain.The brain is protected not only by the rigid skull but also by three membranes, or meninges, that surround the brain. The outer membrane, called the dura mater (literally, "hard mother"), is tough and fibrous. The intermediate membrane, called the arachnoid (cobweb), is a thin, web-like tissue. The inner covering, called the pia mater ("tender mother"), is a delicate membrane that is moulded to the surface of the brain.

Between the pia mater and the arachnoid is a clear, colourless liquid called the cerebrospinal fluid. The fluid surrounds the brain and spinal cord and flows through ventricles, or

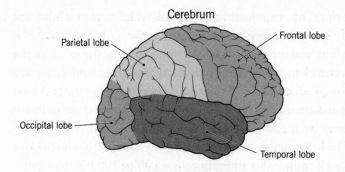

Cerebrum

Parietal lobe

Frontal lobe

Occipital lobe

Temporal lobe

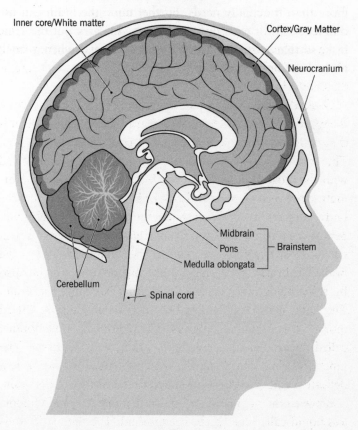

Inner core/White matter

Cortex/Gray Matter

Neurocranium

Midbrain

Pons

Brainstem

Medulla oblongata

Cerebellum

Spinal cord

Figure 1.1   Basic diagram of the brain

cavities. The brain floats in this fluid, which acts as a lubricant and shock absorber, protecting the sensitive brain from trauma. The cerebrospinal fluid also supplements the work of the bloodstream, bringing nutrients to the brain and removing waste products. This duplication of effort between the blood and the cerebrospinal fluid protects the brain from contamination by waste products carried by the blood.

The brain is also protected simply by the fact that it contains such a large number of brain cells – 100 to 200 billion – many more than it actually needs. Furthermore, the brain can use different cells or even different regions to carry out the same task, so that normal functioning need not end when a small part of the brain is damaged.

BASIC STRUCTURE OF THE BRAIN – AN OVERVIEW

Knowing how the brain is put together is essential for understanding important functions such as learning and memory, for understanding the brain's response to injury, and for shedding light on brain disorders such as schizophrenia and intellectual disability, which are thought to result, in part, from a failure to construct proper connections during development.

The brain is encased in a bony vault, called the neurocranium (the uppermost part of the skull), while the long, cylindrical spinal cord lies in the vertebral canal, formed by successive vertebrae connected by dense ligaments. The brain is made up of three main parts: the forebrain, midbrain, and hindbrain. Among the important structures within these areas are 1) the **cerebrum**, which is made up of the left and right hemispheres, 2) the **brainstem**, which attaches to the spinal cord, and 3) the **cerebellum**, which is located just below and behind the cerebral hemispheres. The brain's surface – the cerebral cortex – is convoluted, or wrinkled, which helps it fit inside the skull.

Cerebrum

The cerebrum is the largest and most highly developed part of the brain and is where complex functions such as action and thought take place.

The cerebrum is divided into four sections, or lobes:

1) the **frontal lobe**, which controls cognition, including speech, planning, and problem solving

2) the **parietal lobe**, which controls sensation, such as touch, pressure, and judging size and shape

3) the **temporal lobe**, which mediates visual and verbal memory, and smell

4) the **occipital lobe**, which controls visual reception and recognition of shapes and colours.

Deep "valleys" in the cortex surface, known as sulci, or fissures, form boundaries between the lobes. The central sulcus (also known as the fissure of Rolando) separates the frontal and parietal lobes, and the lateral sulcus (fissure of Sylvius) borders the temporal, frontal, and parietal lobes.

Symmetrical in structure, the cerebrum is divided into the left and right hemispheres. In most people, the left hemisphere is responsible for functions such as creativity, and the right hemisphere is responsible for functions including logic and spatial perception. The left hemisphere controls the movement of the right half of the body, and the right hemisphere controls the movement of the left half of the body. This is because the nerve fibres that send messages to the body cross over in the medulla, part of the brainstem.

Diencephalon

The diencephalon lies beneath the cerebrum, on top of the brainstem. It includes the epithalamus, thalamus, hypothalamus, and subthalamus and has various functions, such as acting as a relay system between incoming sensory input and other areas of the brain, and as a site for interaction between the central nervous and endocrine systems. It also has a role in the limbic system, which is located below the cerebrum and in front of the cerebellum, and is responsible for hereditary and species-specific traits, such as emotion and memory, and which plays a part in regulating basic body functions.

Brainstem

Found at the base of the brain and beneath the deep structures of the cerebral hemispheres and cervical spinal cord, the brainstem (see Figure 1.5) is responsible for basic functions that are essential to life, such as heartbeat, blood pressure, and breathing. The brainstem comprises the midbrain, pons, and medulla oblongata.

Midbrain

The midbrain controls the visual and auditory systems, as well as eye movement, and assists in the control of posture.

Pons

The pons relays sensory information between the cerebellum and cerebrum and controls such functions as arousal and respiration.

Medulla oblongata

The medulla oblongata controls essential involuntary bodily functions, such as breathing, blood pressure, and heart rate.

Reticular activating system

The reticular activating system (RAS) is a group of nerve cells that lies between the midbrain and medulla oblongata and extends up into the thalamus. It is concerned with consciousness and alertness.

Cerebellum

Similar to the cerebrum, the cerebellum has two hemispheres. It acts as a regulator and coordinator of nerve impulses between the brain and muscles and controls functions, such as limb movements, posture, and balance. It plays no role in sensory perception but instead functions like a computer, providing a quick and clear response to sensory signals.

The Brain in Detail

NAVIGATING THE BRAIN
Scientists use a variety of terms to describe the position of anatomical features within the body:

medial	middle
lateral	side
caudal	below and behind
superior	above, on top
inferior	below, underneath
ventral	lower
contralateral	opposite

The brain consists of three major divisions: the forebrain, consisting of the massive paired hemispheres of the cerebrum, and the diencephalon, which includes the thalamus, hypothalamus, epithalamus, and subthalamus; the midbrain; and the hindbrain, consisting of the pons, the medulla oblongata, and the cerebellum (see Figure 1.1).

Cerebrum

From an evolutionary point of view, the cerebrum is the "youngest", yet most highly developed part of the brain. It is involved with the more complex functions and is generally considered to be the structure that most separates humans from other animals. The two hemispheres consist of an inner core (known as "white matter") and a heavily wrinkled outer layer, or cortex, commonly referred to as "gray matter". The cortex is about 0.1 inch (0.3 cm) thick and is wrinkled into ridges (called gyri) and furrows (called sulci). The deepest

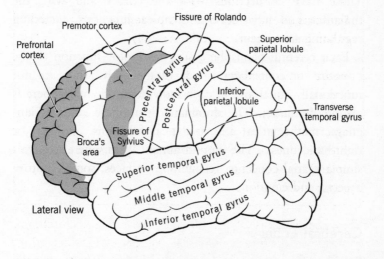

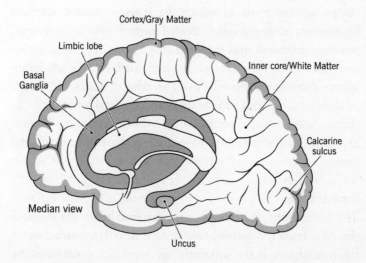

Figure 1.2 Cerebrum

furrows are called fissures. Deep inside the hemispheres are further masses of gray matter called the basal ganglia. These have connections with the cortex and with the thalamus and have several complex functions, including regulating movement.

Each cerebral hemisphere supplies motor function to the opposite, or contralateral, side of the body. They are not functionally equal and in each individual, one hemisphere is always dominant. The dominant hemisphere controls language, mathematical and analytical functions, and left- or right-handedness. The non-dominant hemisphere controls simple spatial concepts, recognition of faces, some auditory aspects, and emotion.

Cerebral Lobes

Sulci and gyri form a more or less constant pattern over the cortex, on the basis of which the surface of each cerebral hemisphere is commonly divided into four lobes: frontal; parietal; temporal; and occipital (see Figure 1.1). Two major sulci located on the lateral, or side, surface of each hemisphere distinguish these lobes. The central sulcus, or fissure of Rolando, separates the frontal and parietal lobes, and the deeper lateral sulcus, or fissure of Sylvius, forms the boundary between the temporal lobe and the frontal and parietal lobes.

Frontal Lobe
The frontal lobe, the largest of the cerebral lobes, lies towards the nose from the central sulcus. One important structure in the frontal lobe is the precentral gyrus, which constitutes the primary motor (movement) region of the brain. When parts of the gyrus are electrically stimulated in conscious patients

(under local anaesthesia), they produce localized movements on the opposite side of the body that are interpreted by the patients as voluntary. Injury to parts of the precentral gyrus results in paralysis on the opposite half of the body.

In front of this region is an area called the premotor cortex, where complex movements are orchestrated. Still farther forward is the prefrontal cortex, which exerts an inhibitory control over actions. Such distinctly human abilities as foreseeing the consequences of an action, exercising self-restraint, and developing moral and ethical standards, depend on the normal functioning of the prefrontal cortex.

Other parts of the frontal lobe (close to the lateral sulcus) constitute Broca's area, a region involved with speech.

Parietal Lobe

The parietal lobe, located behind the central sulcus, is divided into three parts: the postcentral gyrus, and the superior and inferior parietal lobules. The postcentral gyrus receives sensory input from the opposite half of the body. The superior parietal lobule is located below and behind the postcentral gyrus. It is regarded as an association cortex, an area that is not involved in either sensory or motor processing, although part of it may be concerned with motor function. The inferior parietal lobule is involved with the integration of multiple sensory signals.

In both the parietal and frontal lobes, each primary sensory or motor area is close to, or surrounded by, a smaller secondary area. The primary sensory area receives input only from the thalamus, while the secondary sensory area receives input from the thalamus, the primary sensory area, or both. The motor areas receive input from the thalamus as well as the sensory areas of the cerebral cortex.

Temporal Lobe

The temporal lobe, below the lateral sulcus, fills the middle fossa, or hollow area, of the skull. The outer surface of the temporal lobe is an association area (where movement and sensory functions are integrated). Near the margin of the lateral sulcus, two transverse gyri constitute the primary auditory area of the brain. The sensation of hearing is represented here in a tonotopic fashion – that is, with different frequencies represented on different parts of the area. The gyri are surrounded by a less finely tuned secondary auditory area. A protrusion near the lower surface of the temporal lobe, known as the uncus, constitutes a large part of the primary olfactory (smell) area.

Occipital Lobe

The occipital lobe lies below and behind the parieto-occipital sulcus, which joins the calcarine sulcus (the major fissure of the occipital lobe) in a Y-shaped formation. The primary visual area lies on both "banks" of the calcarine sulcus and receives input from the opposite, or contralateral, field of vision via a neural pathway called the optic radiation. The visual field is represented in this area in a retinotopic fashion – that is, different parts of the field of vision correspond to different areas of the banks of the sulcus with upper quadrants of the visual field laid out along the inferior bank of the sulcus and lower quadrants of the visual field represented on the upper quadrant.

The insular, or central, lobe is not visible from the surface of the cerebrum; it is seen in the intact brain only by separating the frontal and parietal lobes from the temporal lobe. The insular lobe is thought to be involved in sensory and motor visceral functions as well as in taste perception.

Limbic Lobe

The limbic lobe is located on the middle margin (limbus, or edge) of the hemisphere. It is composed of adjacent portions of the frontal, parietal, and temporal lobes that surround the corpus callosum (a bundle of neurons that forms the point of union between the two hemispheres of the cerebrum). The limbic lobe is involved with autonomic (involuntary) and related somatic (voluntary) behavioural activities. It receives input from thalamic nuclei that are connected with parts of the hypothalamus and with the hippocampal formation, a primitive cortical structure within the lateral ventricle.

Cerebral Ventricles

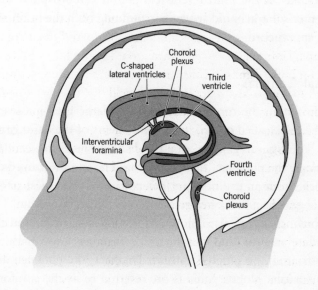

Figure 1.3   Ventricular system

Deep within the white matter of the cerebral hemispheres are cavities filled with cerebrospinal fluid that form the ventricular system. These cavities include a pair of C-shaped lateral ventricles with three "horns" that protrude into the frontal, temporal, and occipital lobes, respectively (see Figure 1.3). Most of the cerebrospinal fluid is produced in the ventricles, and about 70 per cent of it is secreted by the choroid plexus, a collection of blood vessels in the walls of the lateral ventricles. The fluid drains via interventricular foramina, or openings, into a slit-like third ventricle, which, situated along the midline of the brain, separates the symmetrical halves of the thalamus and hypothalamus. From there the fluid passes through the cerebral aqueduct in the midbrain and into the fourth ventricle in the hindbrain. Openings in the fourth ventricle permit cerebrospinal fluid to enter subarachnoid spaces surrounding both the brain and the spinal cord.

Basal Ganglia

Deep within the cerebral hemispheres, large grey masses of nerve cells, called nuclei, form components of the basal ganglia. Four basal ganglia can be distinguished: the caudate nucleus, the putamen, the globus pallidus, and the amygdala, which, from an evolutionary perspective, is the oldest of the basal ganglia and is often referred to as the archistriatum. The globus pallidus is also known as the paleostriatum, and the caudate nucleus and putamen are together known as the neostriatum, or simply striatum. Together, the putamen and the adjacent globus pallidus are referred to as the lentiform nucleus, while the caudate nucleus, putamen, and globus pallidus form the corpus striatum. Because the caudate nucleus and putamen receive varied and diverse inputs from multiple sources, they are regarded as the receptive component of the corpus striatum. Most input originates from regions of the cerebral cortex.

Pathological processes involving the corpus striatum and related nuclei are associated with a variety of specific diseases characterized by abnormal involuntary movements (collectively referred to as dyskinesia) and significant alterations of muscle tone. Parkinson's disease and Huntington's disease are among the more prevalent and each appears related to deficiencies in the synthesis of particular neurotransmitters, or chemical messengers.

The amygdala, located ventral to the corpus striatum in medial parts of the temporal lobe, is an almond-shaped nucleus underlying the uncus. Although it receives olfactory inputs, the amygdala plays no role in olfactory perception.

Diencephalon

Located beneath the cerebrum, on top of the brainstem, the diencephalon's two main components are the thalamus and hypothalamus but it also includes the epithalamus, subthalamus, third ventricle, and the pituitary gland. The thalamus is the major relay centre for all sensations except smell. The hypothalamus controls sex drive, pleasure, pain, hunger, thirst, blood pressure, body temperature, and other functions. The hypothalamus also influences the pituitary, which plays a major part in regulating the functions of endocrine glands, part of the system that releases hormones into the bloodstream and regulates such processes as metabolism, growth, and development. The diencephalon also interconnects with the limbic system, which plays an important role in emotion and memory and in regulating basic body functions.

Limbic System

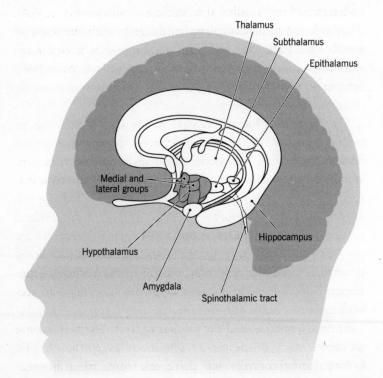

Figure 1.4   Limbic system

The region of the brain that wraps around the brainstem and lies beneath the cerebrum – including such structures as the hippocampus and amygdala (two structures involved with learning and memory), the hypothalamus, and parts of the thalamus – is sometimes considered a functionally related collection of parts called the limbic system (see Figure 1.4). The limbic system is involved particularly with the sense of smell, with memories and with certain complex emotional responses, but it also plays a role in controlling basic body functions such as hunger and thirst.

Thalamus

The thalamus has long been regarded as the key to under-standing the organization of the central nervous system. It is involved in the relay and distribution of most, but not all, sensory and motor signals to specific regions of the cerebral cortex. Sensory signals generated in all types of receptors are projected via complex pathways to specific relay nuclei in the thalamus, where they are segregated and systematically orga-nized. The relay nuclei in turn supply the primary and sec-ondary sensory areas of the cerebral cortex.

The sensory relay nuclei of the thalamus, collectively known as the ventrobasal complex, receive input from the medulla oblongata, from the spinal cord, and from the trigeminal nerve, responsible for sensation in the face. Nerve fibres that terminate in the central core of the ventrobasal complex receive input from deep sensory receptors, whereas fibres projecting on to the outer shell receive input from cutaneous (skin) receptors. This segregation of deep and superficial sensation is preserved in projections of the ventrobasal com-plex to the primary sensory area of the cerebral cortex.

Hypothalamus

The hypothalamus lies below the thalamus in the walls and floor of the third ventricle. It controls major endocrine functions by secreting hormones (i.e. oxytocin and vasopressin) that induce smooth muscle contractions of the reproductive, digestive, and excretory systems. Specific regions of the hypothalamus are also involved with the control of sympathetic and parasympathetic activities (the sympathetic nervous system is widely known for regulating the "fight or flight" response during stressful situations, whilst the parasympathetic system controls functions which do not require immediate action), temperature regulation, food intake, the reproductive cycle, and emotional expression and behaviour.

Epithalamus

The epithalamus is represented mainly by the pineal body or gland, which lies behind the third ventricle. This pea-sized gland synthesizes melatonin and enzymes sensitive to daylight. Rhythmic changes in the activity of the pineal gland in response to daylight suggest that the gland serves as a biological clock.

Subthalamus

The main part of the subthalamus is the subthalamic nucleus, a lens-shaped structure lying behind and to the sides of the hypothalamus. The subthalamic region is traversed by fibres related to the globus pallidus (see section above on basal ganglia). Damage to the subthalamic nucleus produces hemiballismus, a violent movement disorder, in which the limbs are involuntarily flung about.

Brainstem

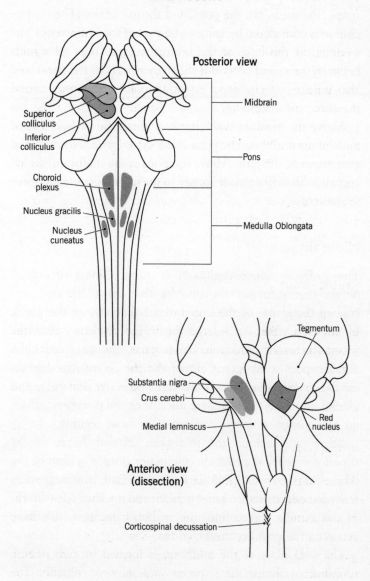

Posterior view

Superior colliculus

Inferior colliculus

Choroid plexus

Nucleus gracilis

Nucleus cuneatus

Midbrain

Pons

Medulla Oblongata

Tegmentum

Substantia nigra

Crus cerebri

Medial lemniscus

Red nucleus

Anterior view (dissection)

Corticospinal decussation

Figure 1.5   Brainstem

The brainstem comprises the medulla oblongata (often called simply the medulla), the pons, and the midbrain. (The diencephalon is considered by some to be part of the brainstem.) The medulla, at the base of the brainstem, transmits all signals between the spinal cord and the higher parts of the brain and also governs mechanisms essential to life: heartbeat, blood pressure, and breathing.

Above the medulla is the pons, a broad, horseshoe-shaped mass of nerve fibres. The pons is associated with sensation and movement of the face. Above the pons is the midbrain, which contains the major motor supply to the muscles controlling eye movements.

Midbrain

The midbrain (mesencephalon) contains a mass of cranial nerves that stimulate the muscles that move the eye, and control the shape of the lens and the diameter of the pupil. In addition, there is a large pigmented nucleus called the substantia nigra, which has two parts, the pars reticulata and the pars compacta. Cells of the pars compacta contain the dark pigment melanin; these cells synthesize dopamine and extend into either the caudate nucleus or the putamen, affecting the output of neurotransmitters in those regions.

Below and behind the midbrain, crossed fibres of the superior cerebellar peduncle (the major output system of the cerebellum) surround and partially terminate in a large, centrally located structure known as the red nucleus. Most fibres of this bundle project into the thalamic nuclei, which have access to the primary motor cortex.

The roof plate of the midbrain is formed by two paired, rounded swellings, the superior and inferior colliculi. The superior colliculus receives input from the retina and the visual cortex and participates in a variety of visual reflexes, particularly the tracking of objects. The inferior colliculus receives auditory fibres and connects with the auditory relay nucleus of the thalamus.

Reticular Activating System (RAS)

Within the brainstem lies a mass of nerve cells and fibres called the reticular formation. Parts of the reticular formation, along with the hypothalamus and the thalamus, excite regions of the cerebrum and keep them active and alert; consequently, these structures and the pathways by which they communicate with the cerebrum are collectively called the reticular activating system. The system sifts information coming to the brain from the senses and transmits only the significant information to the conscious mind.

Pons

The pons (metencephalon) consists of the pontine nuclei, composed of masses of neurons that lie among larger bundles of nerve fibres running lengthways and crossways.

Fibres originating from neurons in the cerebral cortex terminate upon the pontine nuclei, which in turn project to the opposite hemisphere of the cerebellum. These massive crossed fibres, called crus cerebri, form the bridge that connects each cerebral hemisphere with the opposite half of the cerebellum.

The reticular formation of the pontine tegmentum contains multiple cell groups that influence motor function. It also contains the nuclei of several cranial nerves. The facial nerve and the two components of the vestibulocochlear nerve (also known as the auditory or acoustic nerve), for example, emerge from and enter the brainstem at the junction of the pons, medulla, and cerebellum. In addition, motor nuclei of the trigeminal nerve lie in the upper pons. Long ascending and descending tracts that connect the brain to the spinal cord are located on the periphery of the pons.

Medulla Oblongata

The medulla oblongata (myelencephalon), the lowest part of the brainstem, appears as a conical expansion of the spinal cord. The boundary of both the pons and the medulla is formed by the cerebellum and a membrane containing a cellular layer called the choroid plexus. A cerebrospinal fluid-filled space called the cisterna magna surrounds the medulla and the cerebellum.

At the transition of the medulla to the spinal cord, there are two major crossings of nerve fibres (called decussations). The corticospinal decussation is where signals that provide the basis for voluntary motor function on the opposite side of the body are conveyed. In the other decussation, two groups of ascending sensory fibres of the spinal cord terminate. These masses form a major ascending sensory pathway known as the medial lemniscus, which projects into the sensory relay nuclei of the thalamus.

The medulla contains nuclei associated with several of the cranial nerves. It also contains parts of the trigeminal nuclear complex involved with pain and thermal sense.

Cerebellum

The cerebellum ("little brain") serves as a sort of regulator and coordinator of nerve impulses between the brain and the muscles. It lies behind the pons and medulla oblongata and fills the greater part of the base of the skull. This distinctive part of the brain consists of two paired lateral lobes, or hemispheres, and a midline portion known as the vermis. The cerebellar cortex appears very different from the cerebral cortex in that it consists of small, leaf-like plates called folia. The cerebellum consists of a surface cortex of gray matter and a core of white matter containing four paired intrinsic (i.e. deep) nuclei: the dentate, globose, emboliform, and fastigial. Three paired fibre bundles – the superior, middle, and inferior peduncles – connect the cerebellum with the midbrain, pons, and medulla, respectively.

On an embryological basis, the cerebellum is divided into three parts: (1) the archicerebellum, related primarily to the vestibular (balance) system, (2) the paleocerebellum, or anterior lobe, involved with control of muscle tone, and (3) the neocerebellum, known as the posterior lobe. The neocerebellum is the part most concerned with coordination of voluntary motor function.

The cerebellum thus functions as a kind of computer, providing a quick and clear response to sensory signals. It plays no role in sensory perception, but it exerts profound influences upon equilibrium, muscle tone, and the coordination of voluntary motor function. Because the input and output pathways both cross, damage to the cerebellum will affect coordination on the same side of the body.

2

HOW THE BRAIN WORKS

In 1889 Spanish scientist Santiago Ramón y Cajal suggested that the nervous system was composed of individual units that are structurally independent of one another and whose internal contents do not come into direct contact. According to his hypothesis, now known as the neuron theory, each nerve cell communicates with others through contiguity rather than continuity. That is, communication between adjacent but separate cells must take place across the space and barriers separating them.

It has since been proved that Cajal's theory is not universally true, but his central idea – that communication in the nervous system is largely between independent nerve cells – has remained an accurate guiding principle for all further study. There are two basic cell types within the nervous system: neurons and neuroglial (or glial) cells. The neurons perform the essential tasks of the brain, and the glial cells provide a kind of protective environment for the neurons.

Neurons

In the human brain there are approximately 10 billion neurons. Each neuron has its own identity, expressed by its interactions with other neurons and by its secretions. Each also has its own function, depending on its intrinsic properties and location, as well as its inputs from other select groups of neurons, its capacity to integrate those inputs, and its ability to transmit the information to another select group of neurons.

Neurons are the information carriers of the brain. Each neuron consists of a cell body with branching structures, called dendrites, which extend from the cell body like the branches of a tree. In general, the dendrites receive impulses from neighbouring neurons and transmit them to the cell body of the neuron in which they are embedded. Also projecting from the cell body is a single tube-like fibre, called an axon, with tiny branches at its end. Most axons carry nerve impulses away from the cell body to the dendrites of other neurons. Axons may be only a fraction of an inch in length or they may be as long as several feet.

Anatomy of a Neuron

With few exceptions, most neurons consist of three distinct regions, as shown in Figure 2.1: (1) the cell body, or soma (plural somata); (2) the nerve fibre, or axon; and (3) the receiving processes, or dendrites.

1. The neuron is bound by a plasma membrane, a structure so thin that its fine detail can be revealed only by high-resolution electron microscopy. Each neuron contains a nucleus defining the location of the soma. The nucleus is surrounded by a double membrane, called the nuclear envelope, which fuses at intervals to form pores allowing molecular

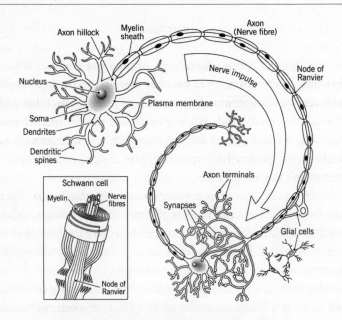

Figure 2.1   Anatomy of a nerve cell

communication with the cytoplasm (the gelatinous fluid that fills most cells). Within the nucleus are the chromosomes, the genetic material of the cell, through which the nucleus controls the synthesis of proteins and the growth and differentiation of the cell into its final form. If the soma dies then so does the neuron.

2. The axon is a cable-like projection that arises from the soma at a region called the axon hillock, or initial segment. This is the region where the plasma membrane generates nerve impulses; the axon conducts these impulses away from the soma or dendrites toward other neurons. Large axons acquire an insulating myelin sheath and are known as myelinated, or medullated, fibres. In the central nervous system, the myelin sheath is formed from glial cells called oligodendrocytes, while in peripheral nerves it is formed from Schwann cells.

The axon transmits nerve impulses by secreting chemical neurotransmitters at junctions called synapses. Synapses are located at the terminal ends (or presynaptic terminals) of axons and lie in close proximity to other neurons and muscle cells. Presynaptic terminals contain synaptic vesicles filled with neurotransmitters and, when observed by light microscope, look like small knobs. Once released, neurotransmitters bind with receptors on the target cell, which can be a dendrite, muscle cell, or gland cell.

3. Dendrites are unmyelinated branches of the neuron that are usually shorter than axons. They form receiving surfaces for the neurotransmitters from other neurons. In many dendrites these surfaces are specialized structures called dendritic spines, which detect changes in electrical current.

The traditional view of dendritic function presumes that only axons conduct nerve impulses and only dendrites receive them, but dendrites can form synapses with dendrites and axons and even somata can receive impulses. Indeed, some neurons have no axon; in these cases nervous transmission is carried out by the dendrites.

Types of neurons
Although all neurons send electrochemical nerve signals, they can differ in structure (in regard to how many axons emanate from the soma) and are found in different parts of the body.

Multipolar or Motor Neurons
Multipolar neurons have just one axon but several dendrites (also called "processes") extending from the soma. They send information away from the central nervous system to muscles or glands.

Sensory or Bipolar Neurons

As the name suggests, sensory neurons send information from the sensory receptors (such as skin, ears, nose, and mouth) toward the central nervous system. They have two processes extending from the soma.

Interneurons or Pseudounipolar neurons

These neurons have two axons rather than an axon and a dendrite. One of these axons extends toward the skin or muscle, and the other extends toward the spinal cord. They send information between motor neurons and sensory neurons.

Glial Cells

Neurons form a minority of the cells in the nervous system. Exceeding them in number by at least ten to one are neuroglial or "glial" cells, which exist in the nervous systems of invertebrates as well as vertebrates.

Glial cells surround the neurons and help regulate the biochemical environment within the brain, provide structural support for neurons, repair the central nervous system after injury, and supply chemicals and other substances that are essential for the healthy functioning of the brain.

The term neuroglia means "nerve glue", and these cells were originally thought to be structural supports for neurons. This idea is still plausible, but other functions of the neuroglia are now generally accepted. Neuroglia can be distinguished from neurons by their lack of axons and by the presence of only one type of process. In addition, they do not form synapses, and they retain the ability to divide throughout their lifespan. Although neurons and neuroglia lie in close proximity to one another, there are no direct junctional specializations, such as gap junctions (intercellular channels that facilitate direct cell-to-cell communication), between the two types. Gap junctions do exist between neuroglial cells, however.

Types of Glial Cells

By staining glial cells with antibodies that bind to specific protein constituents, neurologists have been able to discern two (in some opinions, three) main groups: astrocytes and oligodendrocytes (and sometimes microglia).

Astrocytes

Fibrous astrocytes are found primarily in the white matter of the central nervous system, whereas protoplasmic astrocytes occur in the gray matter. The processes of protoplasmic astrocytes also make contact with capillaries.

Oligodendrocytes

In the white matter of the central nervous system, oligodendrocytes are aligned in rows between the nerve fibres. In the gray matter, they are located close to the somata of neurons. In the peripheral nervous system, neuroglia that are equivalent to oligodendrocytes are called Schwann cells.

Microglia

Microglial cells are small cells with dark cytoplasm and a dark nucleus. It is uncertain whether they are merely damaged neuroglial cells or occur as a separate group in living tissue.

Functions of Glia

Oligodendrocytes and Schwann cells produce the myelin sheath around neuronal axons. A constituent of the axonal surface stimulates Schwann cell proliferation; the type of axon determines whether there is loose or tight myelination of the axon. In tight myelination a glial cell wraps itself like a rolled sheet around a length of axon until the fibre is covered by several layers. Between segments of myelin wrapping are exposed sections called nodes of Ranvier, which are important in the transmission of nerve impulses. Myelinated nerve fibres are found only in vertebrates, leading biologists to conclude that they are an adaptation to transmission over the relatively long distances found in vertebrate nervous systems compared with smaller invertebrates.

Another well-defined role of neuroglial cells is the repair of the central nervous system following injury. Astrocytes divide after injury to the nervous system and occupy the spaces left by injured neurons. The role of oligodendrocytes after injury is unclear, but they may proliferate and form myelin sheaths.

When neurons of the peripheral nervous system are severed, they undergo a process of degeneration followed by regeneration; fibres regenerate in such a way that they return to their original target sites. Schwann cells that remain after nerve degeneration apparently determine the route. This route direction is also performed by astrocytes during the development of the central nervous system. In the developing cerebral cortex and cerebellum of primates, astrocytes project long processes to certain locations, and neurons migrate along these processes to arrive at their final locations. Thus, neuronal organization is brought about to some extent by the neuroglia.

Finally, the environment surrounding neurons in the brain consists of a network of very narrow extracellular clefts. In 1907 Italian biologist Emilio Lugaro suggested that neuroglial cells exchange substances with the extracellular fluid and in this way exert control on the neuronal environment. It has since been shown that glucose, amino acids, and ions – all of which influence neuronal function – are exchanged between the extracellular space and neuroglial cells. After high levels of neuronal activity, for instance, neuroglial cells can regulate potassium ion concentrations and thus maintain normal neuronal function.

How Information is Transmitted

Information is transferred between parts of the brain and between the brain and the spinal cord along chains of interconnected neurons. Each neuron may be connected to 1000 or more other neurons through its branching mesh of dendrites. The transmission of information between neurons, called neurotransmission, is an electrochemical process. Through the pattern of electrochemical exchanges between neurons, the brain receives, analyses, and transmits all the information necessary to carry out its complex functions.

An electrical impulse travels through the cell body of a neuron and down the axon. When it reaches the synapse, it must cross a gap about a millionth of an inch wide (approximately 20 nanometres) to reach the next neuron. The end of the axon contains tiny sacs that hold chemical messengers called neurotransmitters. When the electrical impulse stimulates these sacs, the neurotransmitters are released. The neurotransmitters then diffuse across the gap (or synaptic cleft) and attach to receptor sites on the dendrites of the receiving neuron, sparking an electrical impulse in the receiving neuron. In this manner, a message within the brain is converted, as it moves from one nerve cell to another, from an electrical signal to a chemical signal and back again.

The transfer of information across synapses is a far more complex process, however, than mere nerve stimulation. Every signal sent across a synapse to another cell does not necessarily tell the receiving cell to fire. Quite the opposite – the signal may release a neurotransmitter that tends to prevent the firing of the second cell. If this were not so, the results would be disastrous, because if more than a tiny fraction of the many billions of neurons fired at once, severe and possibly fatal epileptic convulsions would result. A constant interplay between signals that excite the receiving neuron and signals that inhibit it determines whether the neuron will fire.

Subtle variations in the mechanisms of neurotransmission allow the brain to respond to the various demands made on it. Some actions require split-second responses – withdrawing a hand from a hot stove, for example. To relay the information necessary for such a reaction, there are large nerve fibres that can conduct impulses at speeds as high as 330 feet (100 metres) per second. Other activities, such as scholarly pursuits, may require a lifetime of thought. For these kinds of activities, other nerve fibres can be used to conduct signals more slowly – 70 to 100 feet (20 to 30 metres) per second.

Another reason that the brain is so remarkably capable of adapting to the environment is that it has such a large number of chemical messengers with which to work. These include about a dozen neurotransmitters and more than 50 neuropeptides – molecules that serve as neurotransmitters and as modulators in the brain. Considering that any one or more of these neuromessengers can be released, either singly or in combination, by any of the brain's 100 to 200 billion neurons, each with its connections to a thousand or more other neurons, it is clear that the language of the brain is exceedingly complex and the number of possible brain states is inconceivably large.

Action Potential of a Neuron

In the nervous systems of animals at all levels of the evolutionary scale, the signals containing information about a particular stimulus are electrical in nature. In the past, the nerve fibre and its contents were compared to metal wire, while the membrane was compared to insulation around the wire. This comparison was erroneous for a number of reasons. First, the charge carriers in nerves are ions (see below), not electrons, and the density of ions in the axon is much less than that of electrons in a metal wire. Second, the membrane of an axon is not a perfect insulator, so that the movement of current along the axon is not complete. Finally, nerve fibres are smaller than most wires, so that the currents they can carry are limited in amplitude.

Ions are atoms or groups of atoms that gain an electrical charge by losing or acquiring electrons. The important ions in the nervous system are sodium (Na^+) and potassium (K^+), both of which have one positive charge, calcium (Ca^{++}), which has two positive charges, and chloride (Cl^-), which has one negative charge. A positively charged ion is called a cation; a negatively charged ion, an anion. The electrical events that constitute signalling in the nervous system depend upon the distribution of such ions on either side of the nerve membrane. An imbalance in electrical charge across the membrane creates a potential difference, or voltage, that causes current to flow.

In addition, the nerve cell membranes allow some ions to pass but block others, depending on the size of the pores. A membrane with pores allowing the passage of molecules of only a particular size is called a semipermeable membrane. The process of diffusion from an area of high to low concentration across a semipermeable membrane is called osmosis.

Resting Membrane Potential

Three characteristics of the neuron – semipermeability of the membrane, osmotic balance, and electrical charge across the membrane – enable the diffusional and electrical forces to counterbalance and establish an equilibrium, or resting membrane potential. This occurs when the inside of the membrane is more negative than the outside. In most neurons this potential is between –60 and –75 millivolts (mV; or thousandths of a volt; the minus sign indicates that the inner surface is negative). When the inside of the plasma membrane has a negative charge compared to the outside, the neuron is said to be polarized. Any change in resting membrane potential that makes the inside even more negative is called hyperpolarization, while any change that makes it less negative is called depolarization.

The electrical potential across the nerve membrane can be measured by placing one microelectrode within the neuron (usually in the soma) and a second microelectrode in the extracellular fluid. The microelectrode consists of a sharp-tipped, glass capillary tube filled with conducting solution. Upon penetration of the neuron, the potential at the tip of the electrode becomes electrically negative in relation to the outside of the electrode. The value of this negative charge is usually between –60 and –75 mV. This is the membrane potential of the neuron at rest (i.e. when it is not generating a nerve impulse), and for this reason it is called the resting potential.

The resting potential is maintained by the sodium–potassium pump, which steadily discharges more positive charge from the cell than it allows in, and by the properties of potassium ions, which leak out of the cell through its membrane channels faster than sodium ions leak in.

Stimulation

When a physical stimulus, such as touch, taste, or colour, acts on a sensory receptor cell specifically designed to respond to that stimulus, then the energy of the stimulus (e.g. mechanical, chemical, light) is transduced, or transformed, into an electrical response. This response is called the receptor potential, a type of local potential that, when it reaches high enough amplitude, generates the nerve impulse.

Sensory receptors transduce stimuli into electrical responses by activating ion channels in their membranes. (Ion channels are proteins that surround membrane pores and regulate the flow of ions through the pores.) For example, in the stretch receptors of neurons attached to muscle cells, the stretching action of the muscle is thought to put a mechanical stress on protein filaments of the cytoskeleton, which in turn alter the shape of ion channels, inducing them to open and allowing cations to diffuse into the cell. Receptor cells sensitive to chemical and light energy, on the other hand, activate ion channels through the second-messenger system. In this system, stimulated receptor molecules on the surface of the cell membrane catalyse a series of enzymatic reactions within the cytoplasm; these reactions in turn release energy, which activates the ion channels.

Depolarization

By permitting a flux of positive sodium ions into the cell, the opening of ion channels slightly depolarizes the membrane. The extent to which the membrane is depolarized depends upon the extent to which the sodium channels are activated, and this in turn depends upon the strength and duration of the original stimulus at the receptor. If depolarization reaches what is called the threshold potential, it triggers the nerve impulse, or action potential. If it does not reach that amplitude, then the neuron remains at rest, and the local potential, through a process called passive spread, diffuses along the nerve fibre and back out through the membrane.

Peak

Because it varies in amplitude, the local potential is said to be graded. The greater the influx of positive charge – and, consequently, depolarization of the membrane – the higher the grade. Beginning at the resting potential of a neuron (for instance, –75 mV), a local potential can be of any grade up to the threshold potential (for instance, –58 mV). At the threshold, voltage-dependent sodium channels become fully activated. Almost instantly the membrane actually reverses polarity, and the inside acquires a positive charge in relation to the outside.

Repolarization

This reverse polarity constitutes the nerve impulse. It is called the action potential because the positive charge then flows through the cytoplasm, activating sodium channels along the entire length of the nerve fibre. This series of activations, by propagating the action potential along the fibre with virtually no reduction in amplitude, gives the nerve impulse its regenerative property.

Researchers call the nerve impulse an "all-or-none" reaction because there are no gradations between threshold potential and fully activated potential. The neuron is either at rest with a polarized membrane, or it is conducting a nerve impulse at reverse polarization. The reverse polarity of active neurons is measured at about +30 mV.

Sodium Inactivation

As instantaneous as the opening of sodium channels at threshold potential is their closing at the peak of action potential.

This is called sodium inactivation, and it is caused by gates within the channel that are sensitive to depolarization.

Refractory Period

After repolarization there is a period during which a second action potential cannot be initiated, no matter how large a stimulus current is applied to the neuron. This is called the absolute refractory period, and it is followed by a relative refractory period, during which another action potential can be generated, but only by a greater stimulus current than that originally needed. This period is followed by the return of the neuronal properties to the threshold levels originally required for the initiation of action potentials.

Neurotransmission

Neurons send and receive messages from each other and the rest of the body. This is called neurotransmission and occurs in three stages:

1) Neurons release neurostransmitters;
2) The neurons bind to the receptors;
3) This binding passes on the neurotransmitter's message.

Communication between the neurons is made possible by the synapses, and this communication can be either electrical or chemical.

Once an action potential has been generated at the axon hillock, it is conducted along the length of the axon until it reaches the terminals, the finger-like extensions of the neuron that are next to other neurons and muscle cells.

Electrical Transmission

An electrical signal in a neuron occurs because ions (electrically charged particles) move across the neuronal membrane.

In electrical transmission, the ionic current flows directly through channels that couple the cells. This method of transmitting nerve impulses, while far less common than chemical transmission, occurs in the nervous systems of invertebrates and lower vertebrates, as well as in the central nervous systems of some mammals. Transmission takes place through gap junctions, which are protein channels that link the cellular contents of adjacent neurons. Direct diffusion of ions through these junctions allows the action potential to be transmitted with little or no delay or distortion, in effect synchronizing the response of an entire group of neurons. The channels often allow ions to diffuse in both directions, but some gated channels restrict transmission to only one direction.

Chemical Transmission

Chemical transmission occurs at chemical synapses. There are two classic preparations for the study of chemical transmission at the synapse. One is the vertebrate neuromuscular junction, and the other is the giant synapse of the squid *Loligo*. These sites have the advantage of being readily accessible for recording by electrodes – especially the squid synapse, which is large enough that electrodes can be inserted directly into the presynaptic terminal and postsynaptic fibre. In addition, only a single synapse is involved at these sites, whereas a single neuron of the central nervous system may have many synapses with many other neurons, each with a different neurotransmitter.

Neurotransmitters are packed into small, membrane-bound synaptic vesicles. Each vesicle contains thousands of neurotransmitter molecules, and there are thousands of vesicles in each axon terminal. Because the neurotransmitter chemicals are packed into separate, almost identically sized vesicles, their release into the synaptic cleft is said to be quantal – that is, they are expelled in parcels, each vesicle adding its contents incrementally to the contents released from other parcels. This quantal release of neurotransmitter has a critical influence on the electrical potential created in the target, or postsynaptic membrane.

After the neurotransmitter is released from the presynaptic terminal, it diffuses across the synaptic cleft and binds to receptor proteins on the postsynaptic membrane. Some receptors are ion channels that open or close when their molecular configuration is altered by the binding action of the neurotransmitter. Others are membrane proteins that, upon activation, catalyse second-messenger reactions within the postsynaptic cell; these reactions in turn open or close the ion channels. Whether acting upon ion channels directly or indirectly, the neurotransmitter molecules cause a sudden change in the permeability of the membrane to specific ions.

Neuromodulators

Further complicating neurotransmitter action is the presence not only of multiple transmitter substances but also neuromodulators. Neuromodulators are substances that do not directly activate ion-channel receptors but that, acting together with neurotransmitters, enhance the excitatory or inhibitory responses of the receptors. It is often impossible to determine, in the presence of many substances, which are transmitters and which are modulators. In addition to the multiplicity of transmitters and modulators there is a multiplicity of receptors. Some receptors directly open ion channels, while others activate the second-messenger system. Whether they activate channels directly or through a second-messenger system, neurotransmitters are considered to be primary messengers.

Dale's principle

The identification of the substance as a neurotransmitter of the central nervous system was a landmark in neuroscience. The concept is called Dale's principle after Sir Henry Dale, a British physiologist who, in 1935, stated that a neurotransmitter released at one axon terminal of a neuron can be presumed to be released at other axon terminals of the same neuron. (Dale's principle refers only to the presynaptic neuron, as the responses of different postsynaptic receptors to a single neurotransmitter can vary in the same or different neurons.)

The first application of Dale's principle was at the mammalian spinal cord, from which motor neurons send their axons to striated muscles, where the terminals are observed to release acetylcholine. According to Dale's principle, all the branches of a single motor neuron axon should release acetylcholine – including the terminals in the spinal cord. In fact, it was found that some collateral branches leave the motor axons and re-enter the gray matter of the spinal cord, where they synapse on to spinal interneurons. The neurotransmitter released at these terminals is acetylcholine. High concentrations of the acetylcholine-synthesizing enzyme, choline acetyltransferase, and the enzyme for its breakdown, acetylcholinesterase, are also found in motor neuron regions of the spinal cord.

Cholinergic receptors

Cholinergic receptors also exist on the presynaptic terminals of neurons that release acetylcholine, as well as on terminals that release other neurotransmitters. These receptors are called autoreceptors, and they probably regulate the release of neurotransmitters at the terminal. There are two main categories of cholinergic receptor, nicotinic and muscarinic. The nicotinic receptor is a channel protein that, upon binding by acetylcholine, opens to allow diffusion of cations. The muscarinic receptor, on the other hand, is a membrane protein; upon stimulation by neurotransmitter, it causes the opening of ion channels indirectly, through a second messenger. For this reason, the action of a muscarinic synapse is relatively slow. Muscarinic receptors predominate at higher levels of the central nervous system, while nicotinic receptors, which are much faster acting, are more prevalent at neurons of the spinal cord and at neuromuscular junctions in skeletal muscle.

The Brain and the Body

The human nervous system differs from that of other mammals chiefly in the great enlargement and elaboration of the cerebral hemispheres. Much of what is known of the functions of the human brain is derived from observations of the effects of disease or by the results of experimentation on animals, particularly the monkey. Such sources of information are clearly inadequate for the elucidation of the nervous activity underlying many properties of the human brain – particularly speech and mental processes. It is not surprising, therefore, that knowledge of the functions of this uniquely complex system, although rapidly expanding, is far from complete.

In the following account of the functions of the human nervous system, there are numerous references to tracts and to less well-defined connections between different regions of the brain and spinal cord. The identification of these pathways is not always a simple matter; indeed, in humans, many are incompletely known or are simply conjectural.

Axonal Destruction

A great deal of information about the human nervous system has been obtained by observing the effects of axonal destruction. If a nerve fibre is severed, the length of axon farthest from the cell body, or soma, will be deprived of the axonal flow of metabolites and will begin to deteriorate. The myelin sheath will also degenerate, so that, for some months after the injury, breakdown products of myelin will be seen under the microscope with special stains. This method is obviously of limited application in humans, as it requires precise lesions and subsequent examination before the myelin has been completely removed.

The staining of degenerated axons and of the terminals that form synapses with other neurons is also possible through the use of silver impregnation, but the techniques are laborious and results sometimes difficult to interpret. That a damaged neuron should show degenerative changes, however difficult to detect, is not unexpected, but the interdependence of neurons is sometimes shown by transneuronal degeneration. Neurons deprived of major input from axons that have been destroyed may themselves atrophy. This phenomenon is called anterograde degeneration. In retrograde degeneration, similar changes may occur in neurons that have lost the main recipient of their outflow.

These anatomical methods are occasionally applicable to human disease. They can also be used post-mortem when lesions of the central nervous system have been deliberately made – for example, in the surgical treatment of intractable pain. Other techniques can be used only in experiments on animals, but these are not always relevant to humans. For example, normal biochemical constituents labelled with a radioactive isotope can be injected into neurons and then transported the length of the axon, where they can be detected by picking up the radioactivity on an X-ray plate.

An observation technique dependent on retrograde axonal flow has been used extensively to demonstrate the origin of fibre tracts. In this technique, the enzyme peroxidase is taken up by axon terminals and is transported up the axon to the soma, where it can be seen by staining.

Neurotransmitters can be stained in post-mortem human material as well as in animals. Success, however, is dependent on examining relatively fresh or frozen material, and results may be greatly altered in tissues obtained from humans or animals that were previously treated with medications affecting neurotransmission.

Electrical stimulation of a region of the nervous system generates nerve impulses in centres receiving input from the site of stimulation. This method, using microelectrodes, has been widely used in animal studies; however, the precise path followed by the artificially generated impulse may be difficult to establish.

Speech

The question of what the brain does to make the mouth speak or the hand write is still incompletely understood, despite a rapidly growing number of studies by specialists in many sciences, including neurology, psychology, psycholinguistics, neurophysiology, aphasiology, speech pathology, cybernetics, and others. A basic understanding, however, has emerged from such study. In evolution, one of the oldest structures in the brain is the so-called limbic system, which evolved as part of the olfactory (smell) sense. It traverses both hemispheres in a front to back direction, connecting many vitally

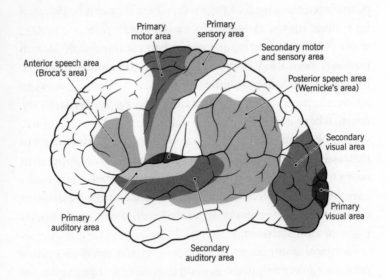

Figure 2.2　Functional areas of the human brain

important brain centres as if it were a basic mainline for the distribution of energy and information. The limbic system involves the reticular activating system (structures in the brainstem), which represents the chief brain mechanism of arousal, such as from sleep or from rest to activity. In humans, all activities of thinking and moving (as expressed by speaking or writing) require the guidance of the cerebral cortex.

Language

In contrast to animals, humans possess several language centres in the dominant brain hemisphere (on the left side in a clearly right-handed person). It was previously believed that left-handers had their dominant hemisphere on the right side, but recent findings tend to show that many left-handed people have the language centres more equally developed in both hemispheres or that the left side of the brain is indeed dominant. Broca's area, in the brain cortex, is involved with motor elaboration of all movements for expressive language. Its destruction through disease or injury causes expressive aphasia, the inability to speak or write (see Figure 2.2). Another cortical region, Wernicke's area, represents receptive speech comprehension. Damage to this area produces receptive aphasia, the inability to understand what is spoken or written as if the patient had never known that language.

Broca's area surrounds and serves to regulate the function of other brain parts that initiate the complex patterns of bodily movement (somatomotor function) necessary for the performance of a given motor act. Swallowing is an inborn reflex (present at birth) in the somatomotor area for mouth, throat, and larynx. From these cells in the motor cortex of the brain emerge fibres that connect eventually with the cranial and spinal nerves that control the muscles of oral speech.

In the opposite direction, fibres from the inner ear have a first relay station in the so-called acoustic nuclei of the brainstem. From here the impulses from the ear ascend, via various regulating relay stations for the acoustic reflexes and directional hearing, to the cortical hearing centre, where the effects of sound stimuli seem to become conscious and understandable. Surrounding this audito-sensory area of initial crude recognition, the inner and outer auditopsychic regions spread over the remainder of the temporal lobe of the brain, where sound signals of all kinds appear to be remembered, comprehended, and fully appreciated. Wernicke's area, in the outer auditopsychic region, appears to be uniquely important for the comprehension of speech sounds.

The integrity of these language areas in the cortex seems insufficient for the smooth production and reception of language. The cortical centres are interconnected with various subcortical areas (deeper within the brain) such as those for emotional integration in the thalamus and for the coordination of movements in the cerebellum (hindbrain).

All creatures regulate their performance instantaneously, comparing it with what it was intended to be through so-called feedback mechanisms involving the nervous system. Auditory feedback through the ear, for example, informs the speaker about the pitch, volume, and inflection of their voice, the accuracy of articulation, the selection of the appropriate words, and other audible features of their utterance. Another feedback system through the proprioceptive sense (represented by sensory structures within muscles, tendons, joints, and other moving parts) provides continual information on the position of these parts. Limitations of these systems curtail the quality of speech as observed in pathological examples (deafness, paralysis, underdevelopment).

Sensory Reception

Ancient philosophers called the human senses "the windows of the soul", and Aristotle enumerated at least five senses – sight, hearing, smell, taste, and touch – and his influence has been so enduring that many people still speak of the five senses as if there were no others. Yet the human skin alone is now regarded as participating in (mediating) a number of different modalities or senses (e.g. hot, cold, pressure, and pain). The modern sensory catalogue also includes a kinesthetic sense (sense organs in muscles, tendons, and joints) and a sense of balance or equilibrium (so-called vestibular organs of the inner ear stimulated by gravity and acceleration). In addition, there are receptors within the circulatory system that are sensitive to carbon dioxide gas in the blood or to changes in blood pressure; and there are receptors in the digestive tract that appear to mediate such experiences as hunger and thirst.

Not all receptors give rise to direct sensory awareness; circulatory (cardiovascular) receptors function largely in reflexes that adjust blood pressure or heart rate without the person being conscious of them. Though perceptible as hunger pangs, feelings of hunger are not exclusively mediated by the gastric (stomach) receptors. Some brain cells may also participate as "hunger" receptors. This is especially true of cells in the lower parts of the brain (such as the hypothalamus) where some cells have been found to be sensitive to changes in blood chemistry (water and other products of digestion) and even to changes in temperature within the brain itself.

Classifying Sensory Structures

One way to classify sensory structures is by the stimuli to which they normally respond; thus, there are photoreceptors (for light), mechanoreceptors (for distortion or bending), thermo-receptors (for heat), chemoreceptors (e.g. for chemical odours), and nociceptors (for painful stimuli). This classification is useful because it makes clear that various sense organs can share common features in the way they convert (transduce) stimulus energy into nerve impulses. Thus, auditory cells and vestibular (balance) receptors in the ear and some receptors in the skin all respond similarly to mechanical displacement (distortion). Because many of the same principles apply to other animals, their receptors can be studied as models of the human senses. In addition, many animals are endowed with specialized receptors that permit them to detect stimuli that humans cannot sense. A snake (the pit viper) boasts a receptor of exquisite sensitivity to "invisible" infrared light; some insects have receptors for ultraviolet light and for pheromones (chemical sex attractants and aphrodisiacs unique to their own species) thereby also exceeding human sensory capabilities.

Basic Features of Sense Organs

Regardless of their specific anatomical form, all sense organs share basic features:

1. They contain receptor cells that are specifically sensitive to one class of stimulus energies, usually within a restricted range of intensity. Such selectivity means that each receptor can be said to have its own "adequate" or proper or normal stimulus, as, for example, light is the adequate stimulus for visual experience. Nevertheless, other energies ("inadequate" stimuli) can also activate the receptor if they are sufficiently intense. Thus, one may "see" pressure when, for example, the thumb is placed on a closed eye and one experiences a bright spot (phosphene) seen in the visual field at a position opposite the touched place.

2. The sensitive mechanism for each modality is often localized in the body at a receiving membrane or surface (such as the retina of the eye) where transducer neurons (sense cells) are to be found. Often the sensory organ incorporates accessory structures to guide the stimulating energy to the receptor cells; thus, the normally transparent cornea and lens within the eye focus light on the retinal sensory neurons. In some cases, blindness can be cured by surgically removing a lens that has grown opaque from cataract to permit light once again to reach the retina. Additional postoperative optical correction in the form of a contact lens or eyeglasses is necessary to compensate for the missing lens. Retinal nerve cells themselves are more or less shielded from non-visual sources of energy by the surrounding structure of the eye; but mild electrical cur-

rents delivered to most sense organs, including the eye, can produce sensory experiences appropriate to the specific organ. The generalized electrical nature of neural function largely accounts for the effectiveness of such currents in evoking a full range of different sensations.

3. The primary transducers or sensory cells in any receptor structure normally connect (synapse) with secondary, ingoing (afferent) nerve cells that carry the nerve impulse along. In some receptors, such as the skin, the individual primary cells possess thread-like structures (axons) that may be yards long, winding from just beneath the skin surface through subcutaneous tissues until they reach the spinal cord. Here each axon from the skin terminates and synapses with the next (second-order) neuron in the chain. By contrast, each primary receptor cell in the eye has a very short axon that is contained entirely in the retina, making synaptic contact with a network of several types of second-order (internuncial) cells, which, in turn, make synaptic contact with third-order neurons called bipolar cells, all still in the retina. The bipolar-cell axons extend beyond the retina, leaving the eyeball to form the optic nerve, which enters the brain to make further synaptic connections. If this visual system is considered as a whole, the retina may be said to be an extended part of the brain on which light can directly fall.

4. From such nerves, still higher order neurons make increasingly complex connections with anatomically separate pathways of the brainstem and deeper parts of the brain (e.g. the thalamus) that eventually end in specific receiving areas in the cerebral cortex (the convoluted outer shell of the brain). Different sensory receiving areas are localized in particular regions of the cortex; e.g. occipital lobes in the back for vision, temporal lobes on the sides for hearing, and parietal lobes towards the top of the brain for tactual function.

Movement

Because of the many differences in the movements used in standing, coughing, laughing, or playing a scale on the piano, it is convenient to think of movements as lower and more automatic or as higher and less automatic. According to this concept, movements are not placed in totally different categories but are regarded as different in degree.

Basic organizations of movement are organized at levels of the central nervous system lower than the cerebral hemispheres – at both the spinal and the brainstem level. Examples of brainstem reflexes are turning of the eyes and head towards a light or sound. The same movements, of course, also can be organized consciously when one decides to turn the head and eyes to look. The cerebral hemispheres themselves can organize certain series of movements, called programmed movements, which need to be performed so rapidly that there is no time for correction of error by local feedback. For this reason the programme is arranged before the movements begin. Examples of such movements are those of a pianist performing a trill or of an athlete hitting a ball.

Automatic Movements

Most of the movements organized by the cerebral cortex are carried out automatically. But when a new series of movements is being learned, or when a movement is difficult, the attributes usually associated with the higher levels of the brain – such as planning, internal speech, remembering, and learning – are used.

The primary motor area is the motor strip of the precentral gyrus. Experiments in monkeys have shown that the motor strip is able to arrange activity of muscles to produce the correct force for the loading conditions of the limbs. To do this, the motor strip continually receives information from the primary sensory area both before and during the movement. Cutaneous areas having the greatest tactile acuity have the largest representation in the primary sensory area; these areas are connected to equally large areas in the primary motor area.

In front of the motor strip is an area known as the premotor cortex or area. When it is stimulated in a monkey, the animal turns its head and eyes as though it is looking in a particular direction. This cortical area, then, organizes the guiding of movements by vision and hearing.

The secondary motor area is at the lower end of the precentral gyrus. It is secondary not only because it was discovered after the primary motor area but also because it does not function in a discrete manner like the primary area. Stimulation of this small area produces movements of large parts of the body. It is also a sensory area, as sensations in the parts of the body being moved are felt during stimulation.

On the medial surface of the hemisphere, in front of the motor strip, is the supplementary motor area. Stimulation of this area can produce vocalization or interrupt speech. Large movements of both sides of the body – often symmetrical movements of the two limbs – also may occur. Stimulation also produces movements of the opposite side of the body, such as raising the upper limb and turning the head and eyes as if looking at something opposite.

In experiments on monkeys, when the animal chooses to respond to one kind of sensation rather than to another, it is the supplementary area that is active rather than the precentral area. In these animals – it is unknown for humans – the fibres descending from the supplementary motor area run to the spinal cord and terminate throughout its whole length. Fibres also are sent to the precentral gyri of both hemispheres, the reticular formation of the pons, the hypothalamus, the mid-brain, and many other masses of cerebral gray matter such as the caudate nucleus and the globus pallidus. The supplementary motor area is upstream from the primary motor area; it initiates movements, whereas the motor strip of the precentral gyrus is part of the apparatus for carrying them out.

Other regions of the cerebral hemisphere from which movements are produced by electrical stimulation are the insula and the surface of the temporal lobe. The insula is a region below the frontal and temporal lobes that, when stimulated, causes movements of the face, larynx, and neck. Stimulation of the anterior end of one temporal lobe causes movements of the head and body towards the other side.

Fibres from the anterior part of the cingulate gyrus are involved in the control of urination and defecation. The organization of these functions also depends on regions anterior to the cingulate gyrus in the medial wall of the frontal lobe. These regions form a part of the limbic lobe, which is responsible, along with their autonomic components, for some emotional states.

Movements Guided by Vision

Movements closely guided by vision have their own pathways. Occipital visual areas send fibres to the pons and from there to the cerebellum. Also just in front of the visual cortex in the parietal lobe are neurons organizing certain types of eye movement. In the monkey, these neurons are at rest during steady gaze, becoming active when the animal turns its eyes to look at something. The fact that the movements constitute a high level of motor behaviour is shown by the activation of these neurons only when the animal is attempting to satisfy an appetite by using its upper limbs and hands; using the limbs for other purposes does not activate them. The neurons are also active when the animal is carrying out the movements of grooming, which also satisfies an innate drive.

One of the main pathways for cortically directed movement of the limbs is the corticospinal tract. This tract developed among animals that used their forelimbs for exploring and affecting the environment as well as for locomotion. It is largest in humans. Fibres of the tract go to various regions of the brainstem and the spinal cord that organize movement. Excitation via the corticospinal tract is then brought to many muscles, all of them presumably working together in a co-ordinated manner. This is achieved by the anatomical arrangement of the motor neurons and by the termination of the corticospinal tract on interneurons, which convey a coordinated pattern of stimulation to the motor neurons.

The corticospinal tract is not merely a pathway to medullary and spinal motor neurons. Activity in this tract can suppress the input from cutaneous areas while facilitating propriocep-tive input. This is probably an important mechanism in the organization of movement. The corticospinal neurons them-selves receive constant input from the cerebellum needed for internal feedback. Much of this input originates in the muscles, joints, and skin of the body parts being moved.

Although a cycle of simple repetitive movements can be organized without sensory feedback, more sophisticated movements require feedback as well as what is called feed-forward control. This is provided by the cerebellum. Many parts of the brain have to be kept informed of movements in order to detect error and continually correct the movement. The cerebellum continuously receives input from the trunk, limbs, eyes, ears, and vestibular apparatus, maintaining in turn a continuous transfer of information to the motor parts of the thalamus and to the cerebral cortex.

As a movement is being prepared, a replica of the instructions is sent to the cerebellum, which sends back its own information to the cerebral cortex. The cortex, meanwhile, sends information about the movement to various afferent neurons that are about to receive information from receptors in the body parts where the movement is about to begin. This comparison between instructions sent and movement performed is a fundamental requirement of all complicated movements. The discharge of impulses from motor to sensory regions is called the corollary discharge. The mechanisms involving the cerebellum do not come to consciousness. There are no sensory consequences of damage to the cerebellum, for the cerebellum is a motor structure.

As series of movements are learned and improved with practice, a replica of the movement is probably retained in the cerebral hemispheres. (The mechanisms of this postulated replica are as yet unknown.) Whenever the learned movements are repeated, they are formed and guided by the replica. This hypothesis of controlling movement by previously practised patterns was developed by German physiologist Erich von Holst. He gave the name "efference" to the totality of motor impulses necessary for a movement, and he proposed that, whenever the efference is produced, it leaves an image of itself somewhere in the central nervous system. He called this image the efference copy. According to von Holst's theory, as the movement is repeated, afferent impulses, called the re-afference, return to the brain from receptors activated by muscular activity. There is then a comparison between the efference copy and the re-afference. When they are identical, the movement is "correct" in relation to its previous performance. When the re-afference differs from the efference copy, corrections have to be made so as to bring the present pattern of movement back to the original image left in the brain.

Diseases that Affect Movement

If the cerebellum is damaged or degenerates, any error between the movement being performed and the efference copy will no longer be corrected, and the postural adjustments sent from the cerebral hemispheres will no longer be implemented. The force and extent of movements also will be abnormal, the movement going too far or not far enough. The various muscles may not come into play at the right time, and there will be a disturbance in the relationship of opposing muscles, so that the accurate arrival on target will be replaced by oscillation.

Most of what is known about the contribution of the basal ganglia has been obtained from studying abnormal conditions that occur when these nuclei are affected by disease. In Parkinson's disease there is a loss of the pigmented neurons of the substantia nigra, which release the neurotransmitter dopamine at synapses in the basal ganglia. Individuals with this disease have a certain type of muscle stiffness called rigidity, a typical tremor, flexed posture, and difficulty in maintaining equilibrium. They have difficulty in initiating movements, including walking, and they cannot put adequate force into fast movements. They have particular difficulty in changing from one movement to its opposite, in carrying out two movements simultaneously, and in stopping one move-ment while starting another.

The organization of posture, which is based on vestibular, proprioceptive, and visual input to the globus pallidus, is severely damaged when this region of the basal ganglia degen-erates. Because a changing posture of the various parts of the body is a prerequisite of every movement, degeneration of this region upsets all movement. Visual reflexes contributing to motion also act through the globus pallidus. A patient may be unable to go forward if they have to pass through a narrow door; another may not be able to do so if they have to go into a wide expanse such as a field.

Perception

To the biologist, the life of animals (including that of humans) consists of seeking stimulation and responding appropriately. A reflex occurs before an individual knows what has happened – for example, what made them lift a foot or drop an object. It is biologically correct to be alarmed before one knows the reason. It is only after the immediate and automatic response that the cerebral cortex is involved and conscious perception begins.

Perception comes between simple sensation and complex cognitional behaviour. It is so automatic that people hardly realize that seeing what they see and hearing what they hear is only an interpretation. Each act of perception is a hypothesis based on prior experience; the world is made up of things people expect to see, hear, or smell, and any new sensory event is perceived in relation to what they already know. People perceive trees, not brown upright masses and blotches of green. Once one has learned to understand speech, it is all but impossible to hear words as sibilants and diphthongs, sounds of lower and higher frequencies. In other words, recognizing a thing entails knowing its total shape or pattern. This is usually called by its German name, Gestalt.

As well as perception of the external environment, there is perception of oneself. Information about one's position in space, for example, comes from vision, from vestibular receptors, and from somatic receptors in the skin and deep tissues. This information is collected in the vestibular nuclei and passed on to the thalamus. From there it is relayed to the central gyri and the parietal region of the cerebral cortex, where it becomes conscious perception.

Perceptual Process

Relations found between various types of stimulation (e.g. light waves and sound waves) and their associated percepts suggest inferences that can be made about the properties of the perceptual process; theories of perceiving can then be developed on the basis of these inferences. Because the perceptual process is not itself public or directly observable (except to the perceiver themselves, whose percepts are given directly in experience), the validity of perceptual theories can be checked only indirectly. That is, predictions derived from theory are compared with appropriate empirical data, quite often through experimental research.

Historically, systematic thought about perceiving was the province of philosophy. Indeed, perceiving remains of interest to philosophers, and many issues about the process that were originally raised by philosophers are still of current concern. As a scientific enterprise, however, the investigation of perception has especially developed as part of the larger discipline of psychology.

Philosophical interest in perception stems largely from questions about the sources and validity of what is called human knowledge. Epistemologists ask whether a real, physical world exists independently of human experience and, if so, how its properties can be learned and how the truth or accuracy of that experience can be determined. They also ask whether there are innate ideas or whether all experience originates through contact with the physical world, mediated by the sense organs. For the most part, psychology bypasses such questions in favour of problems that can be handled by its special methods. The remnants of such philosophical questions, however, do remain; researchers are still concerned, for example, with the relative contributions of innate and learned factors to the perceptual process.

Such fundamental philosophical assertions as the existence of a physical world, however, are taken for granted among most scientific students of perceiving. Typically, researchers in perception simply accept the apparent physical world particularly as it is described in those branches of physics concerned with electromagnetic energy, optics, and mechanics. The problems they consider relate to the process whereby percepts are formed from the interaction of physical energy (for example, light) with the perceiving organism. Of further interest is the degree of correspondence between percepts and the physical objects to which they ordinarily relate. How accurately, for example, does the visually perceived size of an object match its physical size as measured (e.g. with a yardstick)?

Questions of the latter sort imply that perceptual experiences typically have external referents and that they are meaningfully organized, most often as objects. Meaningful objects, such as trees, faces, books, tables, and dogs, are normally seen rather than separately perceived as the dots, lines, colours, and other elements of which they are composed. In the language of Gestalt psychologists, immediate human experience is of organized wholes (*Gestalten*), not of collections of elements.

Gestalt Theory

A major goal of Gestalt theory in the twentieth century was to specify the brain processes that might account for the organization of perception. Gestalt theorists, chief among them the German–American psychologist and philosopher, the founder of Gestalt theory, Max Wertheimer and the German–American psychologists Kurt Koffka and Wolfgang Köhler, rejected the earlier assumption that perceptual organization was the product of learned relationships (associations), the constituent elements of which were called simple sensations. Although Gestaltists agreed that simple sensations logically could be understood to comprise organized percepts, they argued that percepts themselves were basic to experience. One does not perceive so many discrete dots (as simple sensations), for example; the percept is that of a dotted line.

Without denying that learning can play some role in perception, many theorists took the position that perceptual organization reflects innate properties of the brain itself. Indeed, perception and brain functions were held by Gestaltists to be formally identical, so much so that to study perception is to study the brain. Much contemporary research in perception is directed towards inferring specific features of brain function from such behaviour as the reports (introspections) people give of their sensory experiences. More and more such inferences are gratifyingly being matched with physiological observations of the brain itself.

Many investigators relied heavily on introspective reports, treating them as though they were objective descriptions of public events. Serious doubts were raised in the 1920s about this use of introspection by the American psychologist John B. Watson and others, who argued that it yielded only subjective accounts and that percepts are inevitably private experiences and lack the objectivity commonly required of scientific disciplines. In response to objections about subjectivism, there arose an approach known as behaviourism, which restricts its data to objective descriptions or measurements of the overt behaviour of organisms other than the experimenter themselves. Verbal reports are not excluded from consideration as long as they are treated strictly as public (objective) behaviour and are not interpreted as literal, reliable descriptions of the speaker's private (subjective, introspective) experience. The behaviouristic approach does not rule out the scientific investigation of perception; instead, it modestly relegates perceptual events to the status of inferences. Percepts of others manifestly cannot be observed, though their properties can be inferred from observable behaviour (verbal and non-verbal).

Behaviourism

One legacy of behaviourism in contemporary research on perception is a heavy reliance on very simple responses (often non-verbal), such as the pressing of a button or a lever. One advantage of this Spartan approach is that it can be applied to organisms other than humans and to infants (who also cannot give verbal reports). This restriction does not, however, cut off the researcher from the rich supply of hypotheses about perception that derive from their own introspection. Behaviourism does not proscribe sources of hypotheses; it simply specifies that only objective data are to be used in testing those hypotheses.

Behaviouristic methods for studying perception are apt to call minimally on the complex, subjective, so-called higher mental processes that seem characteristic of adult human beings; they thus tend to dehumanize perceptual theory and research. Thus, when attention is limited to objective stimuli and responses, parallels can readily be drawn between perceiving (by living organisms) and information processing (by such devices as electronic computers). Indeed, it is from this information-processing approach that some of the more intriguing theoretical contributions (e.g. abstract models of perception) are currently being made. It is expected that such practical applications as the development of artificial "eyes" for the blind may emerge from these human–machine analogies. Computer-based machines that can discriminate among visual patterns already have been constructed, such as those that "read" the code numbers on bank cheques.

Difference between Sensing and Perceiving

Many philosophers and psychologists have commonly accepted as fundamental a distinction made on rational grounds between sensing and perceiving (or between sensations and percepts). To demonstrate empirically that sensing and perceiving are indeed different, however, is quite another matter. It is often said, for example, that sensations are simple and that percepts are complex. Yet, only if there is offered some agreed upon (a priori) basis for separating experiences into two categories – sensations and percepts – can experimental procedures demonstrate that the items in one category are "simpler" than those in the other. Clearly, the arbitrary basis for the initial categorization itself cannot be subjected to empirical test.

Problems of verification aside, the simplicity–complexity distinction derives from the assumption that percepts are constructed of simple elements that have been joined through association. Presumably, the trained introspectionist can dissociate the constituent elements of a percept from one another, and in so doing, experience them as simple, raw sensations. Efforts to approach the experience of simple sensations might also be made by presenting very simple, brief, isolated stimuli, e.g. flashes of light.

Another commonly offered basis for distinction is the notion that perceiving is subject to the influence of learning while sensing is not. It might be said that the sensations generated by a particular stimulus will be essentially the same from one time to the next (barring fatigue or other temporary changes in sensitivity), while the resulting percepts may vary considerably, depending on what has been learned between one occasion and the next.

Some psychologists have characterized percepts as typically related to external objects and sensations as more nearly subjective, personal, internally localized experiences. Thus, a spontaneous pain in the finger would be called a sensation; however, if the salient feature of experience is that of a painfully sharp, pointed object, such as a pin located "out there", it would be called a percept.

The above definitional criteria all relate to properties of experience; that is, they are psychological. An alternative way of distinguishing between sensing and perceiving that has become widely accepted is physiological-anatomical rather than psychological. In this case, sensations are identified with neural events occurring immediately beyond the sense organ, whereas percepts are identified with activity farther "upstream" in the nervous system, at the level of the brain. This assignment of anatomical locations to sensory and to perceptual processes seems consistent with psychological criteria. That is, the complexity and variability of percepts (both a product of learning) are attributed to the potential for physiological modification inherent in the vastly complex neural circuitry of the brain.

Perception relies on the special senses – visual, auditory, gustatory, and olfactory. Each begins with receptors grouped together in sensory end organs, where sensory input is organized before it is sent to the brain. A reorganization of impulses occurs at every synapse on a sensory pathway, so that by the time an input arrives at the thalamus, it is far from being the original input that stimulated the receptors.

The afferent (ingoing) parts of the thalamus fall into two divisions: a medial part, which is not sensory, and a ventral and lateral part, which is sensory. Nerve impulses reaching the medial part of the thalamus are derived from the reticular formation. This pathway is for emotional and other rapid reactions such as surprise, alarm, vigilance, and the readiness to react. The lateral part of the thalamus is a station on the way to areas of the cerebral cortex that are specific for each kind of sensation (see Figure 2.2).

The cerebral cortex has three somatosensory areas: primary, secondary, and supplemental. The primary area receives input from the ventrolateral thalamus. The secondary area receives input from the lateral part of the thalamus and also auditory and visual input from the geniculate nuclei. The primary and secondary areas are reciprocally connected. The supplementary area lies just behind the primary area.

The cerebral cortex (and the thalamus as well) is composed of non-specific and specific sensory areas. Most neurons of the specific regions have small receptive fields in the periphery, respond to only one kind of stimulus, and follow the features of stimulation exactly. Most neurons of the non-specific regions have large receptive fields and respond to many kinds of stimuli; many do not exactly reproduce the features of the stimulus.

Although different regions of the body are normally represented by specific parts of the somatosensory regions of the cortex, the parts of the body where afferent impulses arrive are not fixed. For example, the leg area is at the top of the postcentral gyrus, but when there is a painful state in the periphery – sciatica, for example – the leg area of the cortex can enlarge and occupy some of the arm area. Furthermore, injury to the peripheral nerves or brain may alter the sensory map of the cortex. These changes in the cortex and similar changes in anatomical function are referred to as plasticity.

From the somatosensory area, nerve fibres run to other regions of the cortex, traditionally called association areas. It is thought that these areas integrate sensory and motor information and that this integration allows objects to be recognized and located in space. With these regions acting upon all their inputs, the brain is carrying out those aspects of neural activity that are commonly labelled mental. It is not known how or where the brain collects messages from sensory receptors and then arranges them to produce a complete representation of the world and of the individual's place in the world.

Vision

The area of the brain concerned with vision makes up the entire occipital lobe and the hind-most parts of the temporal and parietal lobes (see Figure 2.2). The primary visual area, also called the striate cortex, is part of the occipital lobe and is surrounded by the secondary visual area. The visual cortex is sensitive to the position and orientation of edges, the direction and speed of movement of objects in the visual field, and stereoscopic depth, brightness, and colour; these aspects combine to produce visual perception.

The ganglion neurons of the retina are categorized into three functional types: X-, Y-, and W-cells. X-cells have small peripheral fields and are necessary for high-resolution vision. Y-cells are the largest of the three cells, have large peripheral fields, and respond to fast movement. W-cells are the smallest of the three cells, have large peripheral fields, and are sensitive to directional movement. In the retina, 50 to 55 per cent of ganglion cells are W-type, 40 per cent are X-type, and five to 10 per cent are Y-type.

As constituent fibres of the optic nerves and optic tracts, X- and Y-cells connect to the lateral geniculate nucleus of the thalamus, while W-cells connect primarily to the superior colliculus of the midbrain. From these regions, input from the X-cells travels mainly to the primary visual area, that from the Y-cells to the secondary visual area, and that from the W-cells to the area surrounding the secondary area. The collicular pathway serves movement detection and direction of gaze. The tract from the lateral geniculate nucleus is the pathway for visual acuity.

The primary area sends fibres back to the lateral geniculate nucleus, the superior colliculus, and the pupillary reflex centre for feedback control of input to the visual areas. It also sends fibres to the secondary area and to the visual area of the temporal lobe. The secondary area sends fibres to the temporal and parietal lobes. Also, fibres cross from visual areas of one cerebral hemisphere to the other in the corpus callosum. This link allows neurons of the two hemispheres with similar visual fields to have direct contact with each other.

Neurons of the striate cortex may form the first step in appreciation of orientation of objects in the visual field. It is thought, however, that excitation of cortical neurons is insufficient to account for orientation and that inhibition of other neurons in the visual cortex is also necessary. Whatever the mechanism, experiments on cats and monkeys have shown that individual neurons are activated by lines at different angles – for example, at 90° to the horizontal or at an angle of 45°.

Most neurons of the deeper layers of striate cortex are movement analysers. Some are direction analysers, activated by a line or an edge moving in one direction and silenced when it changes direction (the changed direction then activating other neurons). Some neurons may be excited by a dark line on a bright background and others by a light line on a dark background. Form analysers are located in other regions of the striate cortex; for example, some are activated by rectangles and others by stars. Position neurons respond strongly to a spot located in a certain position and poorly to stimulation of a larger area; others respond only to simultaneous binocular stimulation. Colour-specific neurons are sensitive to red, green, or blue. Each of these neurons is excited by one colour and inhibited by another.

In the secondary visual area, many neurons respond particularly to the direction of moving objects. Neurons activated by colour are not activated by white light. In the part of this area where there are many neurons responding to colour, the periphery of the visual field is not mapped; this is because the periphery of the retina does not contain colour receptors, called cones. The peripheral field is mapped in an area with neurons that respond to movement – notably in the region of the superior temporal gyrus.

It seems that one function of the pathway from the superior colliculus to the temporal and parietal cortices is as a tracking system, enabling the eyes and head to follow moving objects and keep them in the visual field. The pathway from the geniculate nucleus to the primary visual area may be said to perceive what the object is and also how and in what direction it moves.

Some neurons in the parietal cortex become active when a visual stimulus comes in from the edge of the visual field towards the centre, while others are excited by particular movements of the eyes. Other neurons react with remarkable specificity – for example, only when the visual stimulus approaches from the same direction as a stimulus moving on the skin, or during the act of reaching for an object and tracking it with the hand. These parietal neurons greatly depend on the state of vigilance. In monkeys that are apparently merely waiting for something to happen or that have nothing to which to pay attention, the neurons are inactive or minimally active. But when the animal is looking at a visual target whose change it has to detect to obtain a reward, the parietal neurons become active.

A great number of neurons of the middle temporal area are sensitive to the direction of movement of a visual stimulus and to the size of an object. Neurons involved in perceiving shape and colour are located in the inferior temporal area. The neurons of the superior temporal polysensory area respond best to moving stimuli – in particular to movements away from the centre of the visual field. Both these areas are involved with the incorporation of visual stimuli and movement.

Hearing

Much of the knowledge of the neurological organization of hearing has been acquired from studies on the bat, an animal that relies on acoustic information for its survival.

In the cochlea (the specialized auditory end organ of the inner ear), the frequency of a pure tone is reported by the location of the reacting neurons in the basilar membrane, and the loudness of the sound is reported by the rate of discharge of nerve impulses. From the cochlea, the auditory input is sent to many auditory nuclei. From there, the auditory input is sent to the medial geniculate nucleus and the inferior colliculus, as with the relay stations of the retina. The auditory input finally goes to the primary and secondary auditory areas of the temporal lobes (see Figure 2.2).

The auditory cortex provides the temporal and spatial frames of reference for the auditory data that it receives. In other words, it is sensitive to aspects of sound more complex than frequency. For instance, there are neurons that react only when a sound starts or stops. Other neurons are sensitive only to particular durations of sound. When a sound is repeated many times, some neurons respond, while others stop responding. Some neurons are sensitive to differences in the intensity and timing of sounds reaching the ears. Certain neurons that never respond to a note of constant frequency respond when the frequency falls or rises. Others respond to the rate of change of frequency, providing information on whether distance from the source of a sound is increasing or decreasing. Some neurons respond to the ear on the same side, others to the opposite ear, and yet others to both ears.

Pain

Touch is the sense by which we determine the characteristics of objects: size, shape, and texture. We do this through touch receptors in the skin. In hairy skin areas, some receptors consist of concentrations of sensory nerve cell endings wrapped around the base of hairs. The nerve endings are very sensitive, being triggered by slight movement of the hairs. Other receptors are more common in non-hairy areas, such as the lips and fingertips, and consist of nerve cell endings that may be free or surrounded by bulb-like structures. Signals from touch receptors pass via sensory nerves to the spinal cord, where they synapse (make contact) and then travel to the thalamus and sensory cortex. The transmission of this information is highly topographic, meaning that the body is represented in an orderly fashion at different levels of the nervous system. Larger areas of the cortex are devoted to sensations from the hands and lips; much smaller cortical regions represent less sensitive parts of the body. Different parts of the body vary in their sensitivity to touch discrimination and painful stimuli.

Pain is a complex experience consisting of a physiological and emotional response to a noxious stimulus. It is a warning mechanism that protects an organism by influencing it to withdraw from harmful stimuli; it is primarily associated with injury or the threat of injury.

Measuring Pain

Pain is subjective and difficult to quantify because it has both an affective (emotional) and a sensory component. Although the neuroanatomical basis of pain reception develops before birth, individual pain responses are learned in early childhood and are affected by social, cultural, psychological, cognitive, and genetic factors, among others. These factors account for differences in pain tolerance among humans. Athletes, for example, may be able to withstand or ignore pain while engaged in a sport, and certain religious practices may require participants to endure pain that seems intolerable to most people.

Hence, we can see that an important function of pain is to alert the body to potential damage (nociception). The pain sensation, however, is only one part of the nociceptive response, which may include an increase in blood pressure, an increase in heart rate, and a reflexive withdrawal from the noxious stimulus. Acute pain can arise from breaking a bone or touching a hot surface. During acute pain an immediate, intense feeling of short duration, sometimes described as a sharp, pricking sensation, is followed by a dull, throbbing sensation. Chronic pain, which is often associated with diseases such as cancer or arthritis, is more difficult to locate and treat. If pain cannot be alleviated, psychological factors such as depression and anxiety can intensify the condition.

Many new insights into the pain experience are coming from studies in which modern imaging tools are used to monitor brain activity when pain is experienced. One finding is that there is no single area in the brain where pain is generated; rather, there are both emotional and sensory components.

When people are hypnotized so that a painful stimulus is not experienced as unpleasant, activity in only some areas of the brain is suppressed. As such techniques for brain study improve, it should be possible to monitor better the changes in the brain that occur in people with persistent pain and to evaluate more effectively the different analgesic drugs being developed.

3

LOOKING AT THE BRAIN

The recent advances in understanding the brain are due to the development of techniques that allow scientists to directly monitor neurons throughout the body. Electrophysiological recordings trace brain electrical activity in response to a specific external stimulus. Electrodes are placed in specific parts of the brain – depending on which sensory system is being tested – and make recordings that are then processed by a computer. The computer makes an analysis based on the time lapse between stimulus and response. It then extracts this information from background activity. Other recent advances have been made in gene diagnosis, and each of these diagnostic methods shall be reviewed in this chapter.

Study of the Brain

Study of the brain is a branch of neurology, the medical specialty concerned with the nervous system and its functional or organic disorders. Neurologists diagnose and treat diseases and disorders of the brain, spinal cord, and nerves.

The first scientific studies of nerve function in animals were performed in the early eighteenth century by English physiologist Stephen Hales and Scottish physiologist Robert Whytt. Knowledge was gained in the late nineteenth century about the causes of aphasia, epilepsy, and motor problems arising from brain damage. French neurologist Jean-Martin Charcot and English neurologist William Gowers described and classified many diseases of the nervous system. The mapping of the functional areas of the brain through selective electrical stimulation also began in the nineteenth century. Despite these contributions, however, most knowledge of the brain and nervous functions came from studies in animals and from the microscopic analysis of nerve cells.

The electroencephalogram (EEG), which records electrical brain activity, was invented in the 1920s by Hans Berger. Development of the EEG, analysis of cerebrospinal fluid obtained by lumbar puncture (spinal tap), and development of cerebral angiography allowed neurologists to increase the precision of their diagnoses and develop specific therapies and rehabilitative measures. Further aiding the diagnosis and treatment of brain disorders were the development of computerized axial tomography (CAT) scanning (today known as computed tomography, CT) in the early 1970s and magnetic resonance imaging (MRI) in the 1980s, both of which yielded detailed, noninvasive views of the inside of the brain. The identification of chemical agents in the central nervous system and the elucidation of their roles in transmitting and blocking nerve impulses have led to the introduction of a wide array of medications that can correct or alleviate various neurological disorders including Parkinson's disease, multiple sclerosis, and epilepsy. Neurosurgery, a medical specialty related to neurology, has also benefited from CT scanning and other increasingly precise methods of locating lesions and other abnormalities in nervous tissues. These and other techniques are described in more detail below.

Cerebral Angiography

Cerebral angiography is X-ray examination of intracranial blood vessels after injection of radiopaque dye into the neck (carotid) artery. Whether arteries or veins are visualized depends on how long the film is exposed after the injection. Cerebral angiography detects solid lesions by showing blood-vessel deformities or displacement. It reveals areas without blood vessels, where cysts and abscesses of the brain are likely to exist. The process was introduced and developed between 1927 and 1937 by António Egas Moniz.

Electroencephalography

Electroncephalography is a technique for recording and interpreting the electrical activity of the brain. The nerve cells of the brain generate electrical impulses that fluctuate rhythmically in distinct patterns. In 1929 Hans Berger of Germany developed an electroencephalograph, an instrument that measures and records these brain wave patterns. The recording produced by such an instrument is called an electroencephalogram, commonly abbreviated to EEG.

To make an EEG, electrodes are placed in pairs on the scalp. Each pair of electrodes transmits a signal to one of several recording channels of the electroencephalograph. This signal consists of the difference in the voltage between the pair. The rhythmic fluctuation of this potential difference is shown as peaks and troughs on a line graph by the recording channel. The EEG of a normal adult in a fully conscious but relaxed state is made up of regularly recurring oscillating waves known as alpha waves. When a person is excited or startled, the alpha waves are replaced by low-voltage, rapid, irregular waves. During sleep, the brain waves become extremely slow. This is also the case when someone is in a deep coma. Other abnormal conditions are associated with particular EEG patterns. Irregular slow waves known as delta waves, for example, arise from the vicinity of a localized area of brain damage.

Electroencephalography provides a means of studying how the brain works and of tracing connections between one part of the central nervous system and another. Its effectiveness as a research tool, however, is limited because it records only a small sample of electrical activity from the surface of the brain. Many of the more complex functions of the brain, such as those that underlie emotions and thought, cannot be related closely to EEG patterns. Electroencephalography has proved more useful as a diagnostic aid in cases of serious head injuries, brain tumours, cerebral infections, epilepsy, and various degenerative diseases of the nervous system.

Brain Scanning

Brain scanning encompasses a number of diagnostic methods for detecting intracranial abnormalities.

The oldest of the brain scanning procedures still in use is a simple, relatively non-invasive procedure called isotope scanning. It is based on the tendency of certain radioactive isotopes to concentrate selectively in tumours and blood vessel lesions. The procedure involves the injection of a radioactive isotope (such as technetium-99m or iodine-131) into a blood vessel that supplies the cranial region. As the substance becomes localized within the brain, it decays, emitting gamma rays in the process. The concentration of rays at a given site, as measured by a movable radiation detection device, can reveal the presence, the shape, and often the size of the intracranial abnormality. In many cases, isotope scanning has been replaced by computed tomography (CT).

The CT scan is a procedure in which the brain is X-rayed from many different angles. An X-ray source delivers a series of short pulses of radiation as it and an electronic detector are rotated around the head of the individual being tested. The responses of the detector are fed to a computer that analyses and integrates the X-ray data from the numerous scans to construct a detailed cross-sectional image of the brain. A series of such images enables physicians to locate brain tumours, cerebral abscesses, blood clots, and other disorders that would be difficult to detect with conventional X-ray techniques.

With the development in the early 1970s of the CT scan, computer-based technologies have revolutionized the field of medical diagnosis. One of the more significant new tomographic techniques is nuclear magnetic resonance (NMR) imaging. Like CT, NMR generates images of thin slices of the brain (or other organ under study), but it does so without the hazard of X-rays or other ionizing radiation. In addition NMR can reveal physiological and biochemical, as well as structural, abnormalities. (Although the benefits of NMR are myriad, the technique is not advised for individuals with pacemakers, aneurysm clips, large metallic prostheses, or dependence on iron-containing instruments.) Positron emission tomography (PET) is a computer-based procedure in which a radioactive tracer-labelled compound is introduced into the brain (or other organ under study), and its behaviour tracked. This information, with computer modelling, eventually yields a cross-sectional image of the physiological process under study.

Imaging Techniques

Tomography

Tomography (CAT or CT scanning) is a radiological technique for obtaining clear X-ray images of deep internal structures by focusing on a specific plane within the body. Structures that are obscured by overlying organs and soft tissues that are insufficiently delineated on conventional X-rays can thus be adequately visualized.

The simplest method is linear tomography, in which the X-ray tube is moved in a straight line in one direction while the film moves in the opposite direction. As these shifts occur, the X-ray tube continues to emit radiation so that most structures in the part of the body under examination are blurred by motion. Only those objects lying in a plane coinciding with the pivot point of a line between the tube and the film are in focus. A somewhat more complicated technique, known as multi-directional tomography, produces an even sharper image by moving the film and X-ray tube in a circular or elliptical pattern. As long as both tube and film move in synchrony, a clear image of objects in the focal plane can be produced. A still more complex technique, computed tomography (CT), was developed by Godfrey Hounsfield of Great Britain and Allen Cormack of the USA during the early 1970s. Since then it has become a widely used diagnostic approach. In this procedure a narrow beam of X-rays sweeps across an area of the body and is recorded not on film but by a radiation detector as a pattern of electrical impulses. Data from many such sweeps are integrated by a computer, which uses the radiation absorption figures to assess the density of tissues at thousands of points. The density values appear on a television-like screen as points of varying brightness to produce a detailed cross-sectional image of the internal structure under scrutiny.

Positron emission tomography

Positron emission tomography (PET) is one of the most important techniques for measuring blood flow or energy consumption in the brain. This method of measuring brain function is based on the detection of radioactivity emitted when positrons, positively-charged particles, undergo radioactive decay in the brain. A chemical compound "labelled" with a short-lived, positron-emitting radionuclide of carbon, oxygen, nitrogen, or fluorine is injected into the body. The activity of such a radiopharmaceutical is quantitatively measured throughout the target organs by means of photomultiplier-scintillator detectors. As the radionuclide decays, positrons are annihilated by electrons, giving rise to gamma-rays that are detected simultaneously by the photomultiplier-scintillator combinations positioned on opposite sides of the patient. The data from the detectors are analysed, integrated, and reconstructed by means of a computer to produce images of the organs being scanned.

PET studies have helped scientists understand more about how drugs affect the brain and what happens during learning, when using language, and in certain brain disorders – such as stroke, depression, and Parkinson's disease. Within the next few years, PET could enable scientists to identify the biochemical nature of neurological and mental disorders and determine how well therapy is working in patients. PET has revealed marked changes in the depressed brain. Knowing the location of these changes helps researchers understand the causes of depression and monitor the effectiveness of specific treatments.

Single photon emission computed tomography

Another technique, single photon emission computed tomography (SPECT), is similar to PET, but its pictures are not as detailed. SPECT is much less expensive than PET because the tracers it uses have a longer half-life and do not require an accelerator nearby to produce them.

Magnetic resonance imaging

Providing a high-quality, three-dimensional image of organs and structures inside the body without X-rays or other radiation, magnetic resonance imaging (MRI) is unsurpassed in anatomical detail and may reveal minute changes that occur over time.

MRIs tell scientists when structural abnormalities first appear in the course of a disease, how they affect subsequent development, and precisely how their progression correlates with mental and emotional aspects of a disorder. Because MRI poorly visualizes bone, excellent images of the intracranial and intraspinal contents are produced.

During the 15-minute MRI procedure, a patient lies inside a massive, hollow, cylindrical magnet and is exposed to a powerful, steady magnetic field. Different atoms in the brain resonate to different frequencies of magnetic fields. In MRI, a background magnetic field lines up all the atoms in the brain. A second magnetic field, oriented differently from the background field, is turned on and off many times a second; at certain pulse rates, particular atoms resonate and line up with this second field.

When the second field is turned off, the atoms that were lined up with it swing back to align with the background field. As they swing back, they create a signal that can be picked up and converted into an image. Tissue that contains a lot of water and fat produces a bright image; tissue that contains little or no water, such as bone, appears black.

MRI images can be constructed in any plane, and the

technique is particularly valuable in studying the brain and spinal cord. It reveals the precise extent of tumours rapidly and vividly. And MRI provides early evidence of potential damage from stroke, allowing physicians to administer proper treatments early.

Magnetic resonance spectroscopy

Magnetic resonance spectroscopy (MRS), a technique related to MRI, uses the same machinery but measures the concentration of specific chemicals – such as neurotransmitters – in different parts of the brain instead of blood flow. MRS also holds great promise: by measuring the molecular and metabolic changes that occur in the brain, this technique has already provided new information on brain development and ageing, Alzheimer's disease, schizophrenia, autism, and stroke. Because it is non-invasive, MRS is ideal for studying the natural course of a disease or its response to treatment.

Functional magnetic resonance imaging

Functional magnetic resonance imaging (fMRI) is currently among the most popular neuroimaging techniques. This technique compares brain activity under resting and activated conditions. It combines the high spatial resolution, non-invasive imaging of brain anatomy offered by standard MRI with a strategy for detecting increases in blood oxygen levels when brain activity brings fresh blood to a particular area of the brain. This technique allows for more detailed maps of brain areas underlying human mental activities in health and disease. To date, fMRI has been applied to the study of various functions of the brain, ranging from primary sensory responses to cognitive activities.

Magnetoencephalography

Magnetoencephalography (MEG) is a technique that reveals the source of weak magnetic fields emitted by neurons. An array of cylinder-shaped sensors monitors the magnetic field pattern near the patient's head to determine the position and strength of activity in various regions of the brain. In contrast with other imaging techniques, MEG can characterize rapidly changing patterns of neural activity – down to millisecond resolution – and can provide a quantitative measure of the strength of this activity in individual subjects. Moreover, by presenting stimuli at various rates, scientists can determine how long neural activation is sustained in the diverse brain areas that respond.

One of the most exciting developments in imaging is the combined use of information from fMRI and MEG. The former provides detailed information about the areas of brain activity in a particular task, whereas MEG tells researchers and physicians when certain areas become active. Together, this information leads to a much more precise understanding of how the brain works in health and disease.

Optical Imaging Techniques

Optical imaging relies on shining weak lasers through the skull to visualize brain activity. These techniques are inexpensive and relatively portable. They are also silent and safe: because only extremely weak lasers are used, they can be used even to study young infants. In a technique called near infrared spectroscopy (NIRS), technicians shine lasers through the skull at near-infrared frequencies, which renders the skull transparent.

Blood with oxygen in it absorbs different frequencies of light than blood in which the oxygen has been consumed. By observing how much light is reflected back from the brain at each frequency, researchers can track blood flow. Diffuse optical tomography is then used to create maps of brain activity. A related technique, the event-related optical signal, records how light is scattered in response to cellular changes that arise when neurons fire and potentially can assess very quickly – in well under a second – changes in neural activity.

Lumbar Puncture

Lumbar puncture, also called spinal tap, is the direct aspiration (fluid withdrawal) of cerebrospinal fluid through a hollow needle. The needle is inserted in the lower back, usually between the third and fourth lumbar vertebrae, into the subarachnoid space of the spinal cord where the cerebrospinal fluid is located. Lumbar puncture is generally performed to obtain pressure measurements and to withdraw cerebrospinal fluid for cellular, chemical, and bacteriological examination; to administer spinal anaesthetics or antibiotics; to inject air or a radiopaque or water-soluble contrast medium substance for myelography; or to inject a radioactive substance to assist in the diagnosis of cerebrospinal fluid leak or hydrocephalus.

Gene Diagnosis

The inherited blueprint for all human characteristics, genes consist of short sections of deoxyribonucleic acid (DNA), the long, spiralling, double-helix structure found on the 23 pairs of chromosomes in the nucleus of every human cell. Gene diagnosis techniques have made it possible to find the chromo-somal location of genes responsible for neurological and psychiatric diseases and to identify structural changes in these genes that are responsible for causing disease. This infor-mation is useful for identifying individuals who carry faulty genes and thereby improving diagnosis, for understanding the precise cause of diseases in order to improve methods of prevention and treatment, and for evaluating the malignancy of and susceptibility to certain tumours. Prenatal or carrier tests exist for many of the most prevalent neurological disorders.

Scientists have tracked down the gene on chromosome 4 that goes awry in Huntington's patients. The defect is an expansion of a CAG repeat. CAG is the genetic code for the amino acid glutamine, and the expanded repeat results in a long string of glutamines within the protein. This expan-sion appears to alter the protein's function. Scientists have found that the size of the expanded repeat in an individual is predictive of Huntington's disease. Other neurodegenerative disorders have been attributed to expanded CAG repeats in other genes. The mechanisms by which these expansions cause adult-onset neurodegeneration is the focus of intense research.

Sometimes patients with single-gene disorders are found to have a chromosomal abnormality – a deletion or break in the DNA sequence of the gene – that can lead scientists to a more accurate position of the disease gene. This is the case with some abnormalities found on the X chromosome in patients with Duchenne muscular dystrophy and on chromosome 13 in patients with inherited retinoblastoma, a rare, highly malig-nant childhood eye tumour that can lead to blindness and death.

Gene mapping has led to the localization on chromosome 21 of the gene coding for the beta amyloid precursor protein that is abnormally cut to form the smaller peptide, beta amyloid. It is this peptide that accumulates in the senile plaques that clog the brains of patients with Alzheimer's disease. This discovery shed light on why individuals with Down's syndrome invariably accumulate amyloid deposits: they make too much amyloid due to having three rather than two copies of this gene (trisomy 21). Mutations in this gene have been shown to underlie Alzheimer's disease in a distinct subset of these patients.

Several other genetic factors have been identified in Alzheimer's disease, including genes for two proteins, presenilin 1 and presenilin 2, located on chromosomes 14 and 1, respectively. A risk factor for late-onset Alzheimer's is the gene for the apolipoprotein E protein located on chromosome 19.

Gene mapping has enabled doctors to diagnose fragile X intellectual disability, the most common cause of inherited intellectual disability. Scientists have identified this gene, which is important for neuronal communication. Scientists have also discovered that some people may have a genetic predisposition to disorders such as schizophrenia, bipolar disorder, and alcoholism. Studies have indicated, however, that genetic predisposition, combined with certain environmental factors, together have the greatest influence on the development of these disorders.

Overall, the characterizations of the structure and function of individual genes causing diseases of the brain and nervous system are in the early stages. Factors that determine variations in the genetic expression of a single-gene abnormality – such as what contributes to the early or late start or severity of a disorder – are still largely unknown.

Scientists also are studying the genes in mitochondria, structures found outside the cell nucleus that have their own DNA and are responsible for the production of energy used by the cell. Recently, different mutations in mitochondrial genes were found to cause several rare neurological disorders. Some scientists speculate that an inheritable variation in mitochondrial DNA may play a role in diseases such as Alzheimer's, Parkinson's, and some childhood diseases of the nervous system.

4

WHAT IS PSYCHOLOGY?

The history of psychology is the history of thought about human consciousness and conduct. Psychological theory has its roots in ancient Greek philosophy and has been fed from streams such as epistemology (the philosophy of knowing), metaphysics, religion, and Oriental philosophy.

Over the centuries psychology and physiology became increasingly separated. A split developed between the essentially phenomenological (experiential) and mechanistic (physiological) conceptions of psychology. In general, through the end of the nineteenth century the British and German traditions were phenomenological, while the French and American were mechanistic. The history of psychology from the nineteenth century may be viewed as a debate between schools of systematic thought concerning the mind, such as associationism, structuralism, and functionalism; or alternatively, as a history of experimentation and research in various areas. Twentieth-century psychology began with structuralism, which employed the method of introspection to describe mental events. It then evolved into psychoanalysis, a derivative of psychiatric tradition, and produced behaviourism and Gestalt psychology, which were reactions against structuralism. Humanistic psychology represented a rebellion against the reductionist and deterministic leanings of earlier schools.

By the Second World War, "schools" of psychology had largely faded away, leaving a common pool of psychological knowledge to which theoreticians, researchers, experimenters, and clinicians all contributed. Biological psychology and psychobiology, fields of study combining psychology and physiology, grew in conjunction with these developments.

The word psychology literally means "study of the mind"; the issue of the relationship of mind and body is pervasive in psychology, owing to its derivation from the fields of philosophy and physiology. Psychology is intimately related to the biological and social sciences.

The broad reach of psychology sometimes gives it the appearance of disunity and promotes the lack of a universally accepted theoretical structure. Some of the divisions within psychology are applied fields, while others are more experimental in nature. The various applied fields include clinical; counselling; industrial, engineering, or personnel; consumer; and environmental. The most important of these specialties, clinical psychology, is concerned with the diagnosis and treatment of mental disorders. Industrial psychology is used in employee selection and related contexts in business and industry.

The broad field known as experimental psychology includes specializations in child, educational, social, developmental, physiological, and comparative psychology. Of these, child psychology applies psychological theory and research methods to children; educational psychology is concerned with learning processes and problems associated with the teaching of students; social psychology is concerned with group dynamics and other aspects of human behaviour in its social and cultural setting; and comparative psychology deals with behaviour as it differs from one species of animal to another. The issues studied by psychologists cover a wide spectrum, comprising learning, cognition, intelligence, motivation, emotion, perception, personality, mental disorders, and the study of the extent to which individual differences are inherited or are shaped environmentally, known as behaviour genetics.

Experimental work using humans as subjects involves legal and ethical limitations. Therefore, a significant amount of research is done with animals, with the hope of transferring the knowledge gained concerning psychophysiological or behavioural functioning to humans.

The methods used in human research include observation (sometimes in non-laboratory settings), interviews, psychological testing (also called psychometrics), laboratory experimentation, and statistical analysis. Psychometrics has in fact become a field in its own right, with psychometrists devising new tools for data collection and analysis and new designs for experimental research.

Neuropsychology

Neuroscientists investigate the structure and functions of the cerebral cortex, but the processes involved in thinking are also studied by cognitive psychologists, who group the mental activities known to the neuroscientist as higher cortical functions under the headings cognitive function or human information processing. From this perspective, complex information processing is the hallmark of cognitive function. Cognitive science attempts to identify and define the processes involved in thinking without regard to their physiological basis. The resulting models of cognitive function resemble

flow charts for a computer program more than neural networks – and, indeed, they frequently make use of computer
terminology and analogies.

The discipline of neuropsychology, by studying the relationship between behaviour and brain function, bridges the gap
between neural and cognitive science. Examples of this
bridging role include studies in which cognitive models are
used as conceptual frameworks to help explain the behaviour
of patients who have suffered damage to different parts of the
brain. Thus, damage to the frontal lobes can be conceptualized
as a failure of the "central executive" component of working
memory, and a failure of the "generate" function in another
model of mental imagery would fit with some of the consequences of left parietal lobe damage.

The analysis of changes in behaviour and ability following
damage to the brain is by far the oldest and probably the most
informative method adopted for studying higher cortical functions. Usually these changes take the form of what is known as
a deficit – that is, an impairment of the ability to act or think in
some way. With certain stipulations, one can assume that the
damaged part of the brain is involved in the function that has
been lost. However, people vary considerably in their abilities,
and most brain lesions occur in subjects whose behaviour was
not formally studied before they became ill. Lesions are rarely
precisely congruent with the brain area responsible for a given
function, and their exact location and extent can be difficult to
determine even with modern imaging techniques. Abnormal
behaviour after brain injury, therefore, is often difficult to
attribute to precisely defined damage or dysfunction.

It would also be naive to suppose that a function is represented in a particular brain area just because it is disrupted
after damage to that area. For example, a tennis champion
does not play well with a broken ankle, but this would not lead

one to conclude that the ankle is the centre in which athletic skill resides. Reasonably certain conclusions about brain–behaviour relationships, therefore, can be drawn only if similar well-defined changes occur reliably in a substantial number of patients suffering from similar lesions or disease states.

The most prominent series of observations clearly belonging to modern neuropsychology was made by Paul Broca in the 1860s. He reported the cases of several patients whose speech had been affected following damage to the left frontal lobe and provided autopsy evidence of the location of the lesion. Broca explicitly recognized the left hemisphere's control of language, one of the fundamental phenomena of higher cortical function.

In 1874 the German neurologist Carl Wernicke described a case in which a lesion in a different part of the left hemisphere, the posterior temporal region, affected language in a different way. In contrast to Broca's cases, language comprehension was more affected than language output. This meant that two different aspects of higher cortical function had been found to be localized in different parts of the brain. In the next few decades there was a rapid expansion in the number of cognitive processes studied and tentatively localized.

Wernicke was one of the first to recognize the importance of the interaction between connected brain areas and to view higher cortical function as the build-up of complex mental processes through the coordinated activities of local regions dealing with relatively simple, predominantly sensory-motor functions. In doing so, he opposed the view of the brain as an equipotential organ acting en masse.

Since Wernicke's time, scientific views have swung between the localization and mass-action theories. Major advances in the twentieth century included vast increases in knowledge, the discovery of new ways of studying the anatomy and physiology of the brain, and the introduction of better quantitative methods in the study of behaviour.

Broca's declaration that the left hemisphere is predominantly responsible for language-related behaviour is the clearest and most dramatic example of an asymmetry of function in the human brain. This functional asymmetry is related to hand preference and probably to anatomical differences, although neither relationship is simple.

Evidence from a number of converging sources, notably the high incidence of the language disturbance aphasia after left-but not right-hemisphere damage, indicates that the left hemisphere is dominant for the comprehension and expression of language in close to 99 per cent of right-handed people. At least 60 per cent of left-handed and ambidextrous people also have left-hemisphere language, but up to 30 per cent have predominantly right-hemisphere language. The remainder have language represented to some degree in both hemispheres.

The posterior temporal region of the brain, which is one of the regions responsible for language in the dominant hemisphere, is physically asymmetrical; specifically, the area known as the planum temporale is larger in the left hemisphere in most people. This asymmetry is more common in right-handers, while left-handed individuals are likely to have more nearly symmetrical brains. Reduced anatomical asymmetry has also been found in people with right-hemisphere dominance for speech and in some people with the reading disorder dyslexia. These results point to some relationship between handedness, cerebral dominance for language, anatomical asymmetry in the temporal lobe, and some aspects of language competence. Certainly there is a tendency for right-handedness, left-hemisphere dominance for language, and a larger left planum temporale to occur together. However, there are exceptions;

for example, a few right-handers are right-hemisphere domi-nant for speech, and some right-handers who have left-hemi-sphere speech do not have a larger left planum temporale. In people who are atypical in one of these respects – for example, by being left-handed – the relationship between handedness, cerebral dominance, and anatomical asymmetry is much less consistent. Therefore, language is not invariably located in the hemisphere opposite the dominant hand or in the hemisphere with the larger planum temporale.

Studies of individuals being treated for epilepsy in whom the corpus callosum (the bundle of nerve fibres connecting the two halves of the brain) has been severed, allowing the two hemi-spheres to function largely independently, have revealed that the right hemisphere has more language competence than was thought. These individuals show evidence of comprehension of words presented to the isolated right hemisphere, although that hemisphere is not able to initiate speech. The speech of individuals with a lesion of the right hemisphere may lack normal melodic quality, and they may have difficulty express-ing and understanding such things as emotional overtones. They may also have difficulty appreciating some of the more subtle, connotative aspects of language, such as puns, figures of speech, and jokes. Nevertheless, the dominance of the left hemisphere for language, particularly the syntactic aspects of language and language output, is the clearest example yet discovered of the lateralization of higher cortical function.

The left hemisphere also appears to be more involved than the right in the programming of complex sequences of move-ment and in some aspects of awareness of one's own body. Thus, apraxia is more common after damage to the left hemisphere. In apraxia, the individual has difficulty perform-ing actions involving several movements or the manipulation of objects in an appropriate and skilful way. The difficulty appears to be in programming the motor system to control the sequence of movements required to perform a complex action in the appropriate order and with the appropriate timing.

Confusion of right and left is also found after left-hemisphere damage, making it appear that the left hemisphere is largely responsible for collating somatosensory information into a special awareness of the body called the body image. Finger agnosia is a condition in which the individual does not appear to "know" which finger is which and is unable to indicate which one the examiner touches without the aid of vision. The phenomenon of the phantom limb, whereby patients "feel" sensations in amputated limbs, indicates that the brain's internal representation of the body may persist intact for some time after the loss of a body part. This internal representation appears to be maintained chiefly by the left hemisphere.

The special functions of the right hemisphere were recognized later than those of the left hemisphere, although a case of "imperception" reported by the English neurologist John Hughlings Jackson in 1876 foreshadowed later findings. Jackson's patient, who had a lesion in the posterior part of the right hemisphere, lost her way in familiar surroundings, failed to recognize familiar places and people, and had difficulty in dressing herself – all of which became well-recognized consequences of right-hemisphere damage. The right hemisphere appears to be specialized for some aspects of higher-level visual perception, spatial orientation, and sense of direction, and it probably plays a dominant role in the recognition of objects and faces. The specialization of the right hemisphere, however, is less absolute than that of the left hemisphere in that these skills are less lateralized than language.

There has been considerable speculation as to why the human brain is functionally asymmetrical. Initially, both functional and anatomical asymmetry were thought, like language, to be a uniquely human trait, but less pronounced asymmetries have now been found in lower animals. One theory is that it is necessary to have language represented in a single hemisphere to avoid competition between the hemispheres for control of the muscles involved in speech. Another theory is that it is efficient to have the language system represented in a restricted area on one side of the brain because information needs to be transferred over short distances and fewer connections. A third theory is that the dominance of the left hemisphere over the right hand and skilled movement preceded its dominance over language. According to this view, language subsequently developed in the same hemisphere because language implies speech, which requires precise programming of sequences of movement in the articulatory musculature. None of these theories has been conclusively proved correct or has been generally accepted. Also, there remain some facts that are difficult to explain by any theory. For example, all of the above theories would predict that bilateral and, in some cases, right-hemisphere language representation would be disadvantageous, but this does not seem to be generally true.

Clinical Psychology

Clinical psychology is the branch of psychology concerned with the practical application of research methodologies and findings in the diagnosis and treatment of mental disorders.

Clinical psychologists classify their basic activities under three main headings: assessment (including diagnosis), treatment, and research. In assessment, clinical psychologists administer and interpret psychological tests, either for the purpose of evaluating individuals' relative intelligence or other capabilities or for the purpose of eliciting mental characteristics that will aid in diagnosing a particular mental disorder. The interview, in which the psychologist observes, questions, and interacts with a patient, is another tool of diagnosis.

For purposes of treatment, the clinical psychologist may use any of several types of psychotherapy. Many clinical psychologists take an eclectic approach, drawing on a combination of techniques suited to the client. Clinical psychologists may specialize in behaviour therapy, group therapy, family therapy, or psychoanalysis, among others.

Research is an important field for some clinical psychologists because of their training in experimental research and statistical procedures. Clinical psychologists are thus often crucial participants in studies relating to mental health care.

Clinical psychologists work in a variety of settings, including hospitals, clinics, and corporations, and in private practice. Some specialize in working with mentally or physically handicapped people, prison inmates, drug and alcohol abusers, or geriatric patients. In some settings, a clinical psychologist works in tandem with a psychiatrist and a social worker and is responsible for conducting the team's research. Clinical psychologists also serve the courts in assessing defendants or potential parolees, and others are employed by the armed forces to evaluate or treat service personnel.

The training of clinical psychologists usually includes university-level study of general psychology and some clinical experience. In the USA, New Mexico became the first state to grant psychologists the right to prescribe medications for the treatment of mental disorders. Most clinical psychologists who do not have medical degrees, however, are barred by state laws from prescribing medications.

Experimental Psychology

Experimental psychology is a method of studying psychological phenomena and processes. The experimental method in psychology attempts to account for the activities of animals (including humans) and the functional organization of mental processes by manipulating variables that may give rise to behaviour; it is primarily concerned with discovering laws that describe manipulable relationships. The term generally connotes all areas of psychology that use the experimental method.

These areas include the study of sensation and perception, learning and memory, motivation, and biological psychology. There are experimental branches in many other areas, however, including child psychology, clinical psychology, educational psychology, and social psychology. Usually the experimental psychologist deals with normal, intact organisms; in biological psychology, however, studies are often conducted with organisms modified by surgery, radiation, drug treatment, or long-standing deprivations of various kinds or with organisms that naturally present organic abnormalities or emotional disorders.

Gestalt Psychology

Gestalt psychology is a school of psychology founded in the twentieth century that provided the foundation for the modern study of perception. Gestalt theory emphasizes that the whole of anything is greater than its parts. That is, the attributes of the whole are not deducible from analysis of the parts in isolation. The word *Gestalt* is used in modern German to mean the way a thing has been "placed" or "put together". There is no exact equivalent in English. "Form" and "shape" are the usual translations; in psychology the word is often interpreted as "pattern" or "configuration".

Gestalt theory originated in Austria and Germany as a reaction against the associationist and structural schools' atomistic orientation (an approach which fragmented experience into distinct and unrelated elements). Gestalt studies made use instead of phenomenology, which involves nothing more than the description of direct psychological experience, with no restrictions on what is permissible in the description. Gestalt psychology was in part an attempt to add a humanistic dimension to what was considered a sterile approach to the scientific study of mental life. Gestalt psychology further sought to encompass the qualities of form, meaning, and value that prevailing psychologists had either ignored or presumed to fall outside the boundaries of science.

The publication of Czech-born psychologist Max Wertheimer's *Experimentelle Studien über das Sehen von Bewegung* ("Experimental Studies of the Perception of Movement") in 1912 marks the founding of the Gestalt school. In it, Wertheimer reported the result of a study on apparent movement conducted in Frankfurt, Germany, with psychologists Wolfgang Köhler and Kurt Koffka. Together, these three formed the core of the Gestalt school for the next few decades.

The earliest Gestalt work concerned perception, with particular emphasis on visual perceptual organization as explained by the phenomenon of illusion. In 1912 Wertheimer discovered the phi phenomenon, an optical illusion in which stationary objects shown in rapid succession, transcending the threshold at which they can be perceived separately, appear to move. The explanation of this phenomenon – also known as persistence of vision and experienced when viewing motion pictures – provided strong support for Gestalt principles.

Under the old assumption that sensations of perceptual experience stand in one-to-one relation to physical stimuli, the effect of the phi phenomenon was apparently inexplicable. However, Wertheimer understood that the perceived motion is an emergent experience, not present in the stimuli in isolation but dependent upon the relational characteristics of the stimuli. As the motion is perceived, the observer's nervous system and experience do not passively register the physical input in a piecemeal way. Rather, the neural organization as well as the perceptual experience springs immediately into existence as an entire field with differentiated parts. In later writings this principle was stated as the law of *Prägnanz*, meaning that the neural and perceptual organization of any set of stimuli will form as good a Gestalt, or whole, as the prevailing conditions will allow.

Major elaborations of the new formulation occurred within the next decades. The Gestalt approach was extended to problems in other areas of perception, problem solving, learning, and thinking. Gestalt principles were later applied to motivation, social psychology, and personality and to aesthetics and economic behaviour.

PART 2

MEMORY, INTELLIGENCE, AND THE MIND

5

BEHAVIOUR AND EMOTIONS

Humans have developed innate drives, desires, and emotions, and the ability to remember and learn (in order to survive). These fundamental features of living depend on the entire brain, yet there is one part of the brain that organizes metabolism, growth, sexual differentiation, and the desires and drives necessary to achieve these aspects of life. This is the hypothalamus and a region in front of it comprising the septal and pre-optic areas.

That such basic aspects of life might depend on a small region of the brain was conceived in the 1920s by the Swiss physiologist Walter Rudolph Hess and later amplified by German physiologist Erich von Holst. Hess implanted electrodes in the hypothalamus and in septal and pre-optic nuclei of cats, stimulated them, and observed the animals' behaviour. Finally, he made minute lesions by means of these electrodes and again observed the effects on behaviour. With this technique he showed that certain kinds of behaviour were organized essentially by just a few neurons in these regions of the brain. Later, von Holst stimulated electrodes by remote con-

trol after placing the animals in various biologically meaningful conditions. When such acts result from artificial stimulation of the neurons, the accompanying emotion also occurs,
as do the movements expressing that emotion.

The hypothalamus, in company with the pituitary gland,
controls the emission of hormones, body temperature, blood
pressure and the rate and force of the heartbeat, and water and
electrolyte levels. The maintenance of these and other changing
events within normal limits is called homeostasis; this includes
behaviour aimed at keeping the body in a correct and thus
comfortable environment.

The hypothalamus is also the centre for organizing the
activity of the two parts of the autonomic system, the parasympathetic and the sympathetic. Above the hypothalamus,
regions of the cerebral hemispheres most closely connected to
the parasympathetic regions are the orbital surface of the
frontal lobes, the insula, and the anterior part of the temporal
lobe. The regions most closely connected to the sympathetic
regions are the anterior nucleus of the thalamus, the hippocampus, and the nuclei connected to these structures.

In general, the regions of the cerebral hemispheres that are
closely related to the hypothalamus are those parts that
together constitute the limbic lobe, first considered as a unit
and given its name in 1878 by the French anatomist Paul
Broca. Together with related nuclei, it is usually called the
limbic system, consisting of the cingulate and parahippocampal gyri, the hippocampus, the amygdala, the septal and pre-
optic nuclei, and their various connections.

The autonomic system also involves the hypothalamus in
controlling movement. Emotional expression, which depends
greatly on the sympathetic nervous system, is controlled by
regions of the cerebral hemispheres above the hypothalamus
and by the midbrain below it.

Behaviour

A great deal of human behaviour involves social interaction. Although the whole brain contributes to social activities, certain parts of the cerebral hemispheres are particularly involved. The surgical procedure of leucotomy, cutting through the white matter that connects parts of the frontal lobes with the thalamus, upsets this aspect of behaviour. This procedure, proposed by the Spanish neurologist Egas Moniz, used to be performed for severe depression or obsessional neuroses. After the procedure, patients lacked the usual inhibitions that were socially demanded, appearing to obey the first impulse that occurred to them. They told people what they thought of them without regard for the necessary conventions of civilization.

Which parts of the cerebral hemispheres produce emotion has been learned from patients with epilepsy and from surgical procedures under local anaesthesia in which the brain is electrically stimulated. The limbic lobe, including the hippocampus, is particularly important in producing emotion. Stimulating certain regions of the temporal lobes produces an intense feeling of fear or dread; stimulating nearby regions produces a feeling of isolation and loneliness, other regions a feeling of disgust, and yet others intense sorrow, depression, anxiety, ecstasy, and, occasionally, guilt.

In addition to these regions of the cerebral cortex and the hypothalamus, regions of the thalamus also contribute to the genesis of emotion. The hypothalamus itself does not initiate behaviour; that is done by the cerebral hemispheres.

As seen in Figure 1.1, each cerebral hemisphere supplies motor function to the opposite, or contralateral, side of the body from which it receives sensory input. In other words, the left hemisphere controls the right half of the body, and vice versa. Each hemisphere also receives impulses conveying the senses of touch and vision, largely from the contralateral half of the body, while auditory input comes from both sides. Pathways conveying the senses of smell and taste to the cerebral cortex are ipsilateral (that is, they do not cross to the opposite hemisphere).

In spite of this arrangement, the cerebral hemispheres are not functionally equal. In each individual, one hemisphere is dominant. The dominant hemisphere controls language, mathematical and analytical functions, and handedness. The nondominant hemisphere controls simple spatial concepts, recognition of faces, some auditory aspects, and emotion.

However, there is some evidence that the two hemispheres of the brain are related differently to emotion processes. Early evidence suggested that the right (or dominant) hemisphere may be more adept than the left at discriminating among emotional expressions. Later research using electroencephalography elaborated this initial conclusion, suggesting that the right hemisphere may be more involved in processing negative emotions and the left hemisphere more involved in processing positive emotions.

The neurons of the cerebral cortex constitute the highest level of control in the hierarchy of the nervous system. Consequently, the terms higher cerebral functions and higher cortical functions are used by neurologists and neuroscientists to refer to all conscious mental activity, such as thinking, remembering, and reasoning, and to complex volitional behaviour such as speaking and carrying out purposive movement. The terms also refer to the processing of information in the cerebral cortex, most of which takes place unconsciously.

For example, when certain neurons of the hypothalamus are excited, an individual either becomes aggressive or flees. These two opposite behaviours are together called the defence reaction, or the fight-or-flight response; both are in the repertoire of all vertebrates. The defence reaction is accompanied by strong sympathetic activity. Aggression is also influenced by the production of androgen hormones.

As another example, the fundamental discovery made in 1954 by Canadian researchers James Olds and Peter Milner found that stimulation of certain regions of the brain of the rat acted as a reward in teaching the animals to run mazes and solve problems. The conclusion from such experiments is that stimulation gives the animals pleasure. The discovery has also been confirmed in humans. These regions are called pleasure, or reward, centres. One important centre is in the septal region, and there are reward centres in the hypothalamus and in the temporal lobes of the cerebral hemispheres as well. When the septal region is stimulated in conscious patients undergoing neurosurgery, they experience feelings of pleasure, optimism, euphoria, and happiness.

Regions of the brain also clearly cause rats distress when electrically stimulated; these are called aversive centres. However, the existence of an aversive centre is less certain than that of a reward centre. Electrodes stimulating neurons or neural pathways may cause an animal to have pain, anxiety, fear, or any unpleasant feeling or emotion. These pathways are not necessarily centres that provide punishment in the sense that a reward centre provides pleasure. Therefore, it is not definitely known that connections to aversive centres punish the animal for biologically wrong behaviour, but it is thought that correct behaviour is rewarded by pleasure provided by neurons of the brain.

Emotions

Both biosocial and constructivist theories of emotions ac-
knowledge that an emotion is a complex phenomenon. They
generally agree that an emotion includes physiological functions, expressive behaviour, and subjective experience and that
each of these components is based on activity in the brain and
nervous system. As noted above, some theorists, particularly
those of the constructivist persuasion, hold that an emotion
also involves cognition, an appraisal or cognitive-evaluative
process that triggers the emotion and determines or contributes to the subjective experience of the emotion.

Since the popularization of the James–Lange theory of
emotion, the physiological component of emotion has been
traditionally identified as activity in the autonomic nervous
system and the visceral organs (e.g. the heart and lungs) that it
innervates. However, some contemporary theorists hold that
the neural basis of emotions resides in the central nervous
system and that the autonomic nervous system is recruited by
emotion to fulfil certain functions related to sustaining and
regulating emotion experience and emotion-related behaviour.
Several findings from neuroscience support this idea. Neuroanatomical studies have shown that the central nervous system
structures involved in emotion activation can exert direct
influences on the autonomic nervous system. For example,
efferents from the amygdala to the hypothalamus may influence activity in the autonomic nervous system that is involved
in defensive reactions. Further, there are connections between
pathways innervating facial expression and the autonomic
nervous system. Studies have shown that patterns of activity
in this system vary with the type of emotion being expressed.

The expressive component of emotion includes facial, vocal,
postural, and gestural activity. Expressive behaviour is

mediated by phylogenetically old structures of the brain, which is consistent with the notion that they served survival functions in the course of evolution.

Early studies of the neural basis of emotion expression showed that aggressive behaviour can be elicited from a cat after its neocortex has been removed and suggested that the hypothalamus is a critical subcortical structure mediating aggression. Later research indicated that, rather than the hypothalamus, the central grey region of the midbrain and the substantia nigra may be the key structures mediating aggressive behaviour in animals.

Of the various types of expressive behaviour, facial expression has received the most attention. In human beings and in many non-human primates, patterns of facial movements constitute the chief means of displaying emotion-specific signals. Whereas research has provided much information on the neural basis of emotional behaviours (e.g. aggression) in animals, little is known about the brain structures that control facial expression.

The peripheral pathways of facial emotion expression consist of the seventh and fifth cranial nerves. The seventh, or facial, nerve is the efferent (outward) pathway; it conveys motor messages from the brain to facial muscles. The fifth, or trigeminal, nerve is the afferent (inward) pathway that provides sensory data from movements of facial muscles and skin. According to some theorists, it is the trigeminal nerve that transmits the facial feedback which contributes to the activation and regulation of emotion experience. The impulses for this sensory feedback originate when movement stimulates the mechanoreceptors in facial skin. The skin is richly supplied with such receptors, and the many branches of the trigeminal nerve detect and convey the sensory impulses to the brain.

Conclusion

The emotions are central to the issues of modern times, but perhaps they have been critical to the issues of every era. Poets, prophets, and philosophers of all ages have recognized the significance of emotions in individual life and human affairs, and the meaning of a specific emotion, at least in the context of verbal expression, seems to be timeless. Although art, literature, and philosophy have contributed to the understanding of emotion experiences throughout the ages, modern science has provided a substantial increase in the knowledge of the neurophysiological basis of emotions and their structure and functions.

Research in neuroscience and developmental psychology suggests that emotions can be activated automatically and unconsciously in subcortical pathways. This suggests that humans often experience emotions without reasoning why. Such precognitive information processing may be continuous, and the resulting emotion states may influence the many perceptual-cognitive and behavioural processes (such as perceiving, thinking, judging, remembering, imagining, and coping) that activate emotions through pathways involving the neocortex.

The two recognized types of emotion activation have important implications for the role of emotions in cognition and action. Subcortical, automatic information processing may provide the primitive data for immediate emotional response, whereas higher-order cognitive information processing involving the neocortex yields the evaluations and attributions necessary for the appropriate emotions and coping strategy in a complex situation.

Biosocial and constructivist theories agree that perception, thought, imagery, and memory are important causes of emotions. They also agree that once emotion is activated, emotion and cognition influence each other. How people feel affects what they perceive, think, and do, and vice versa.

Emotions have physiological, expressive, and experiential components, and each component can be studied in terms of its structure and functions. The physiological component influences the intensity and duration of felt emotion, expressions serve communicative and sociomotivational functions, and emotion experiences (feeling states) influence cognition and action.

Research has shown that certain emotion expressions are innate and universal and have significant functions in infant development and in infant–parent relations and that there are stable individual differences in emotion expressiveness. Emotion states influence what people perceive, learn, and remember, and they are involved in the development of empathic, altruistic, and moral behaviour and in basic personality traits.

6

MEMORY AND LANGUAGE

Memory

Memory is the encoding, storage, and retrieval in the human mind of past experiences. Memory is necessary for the performance of many cognitive tasks. Working, or short-term, memory is the memory one uses, for example, to remember a telephone number after looking it up in a directory and while dialling. In order to understand this sentence, for example, a reader must maintain the first half of the sentence in working memory while reading the second half. The capacity of working memory is limited, and it decreases if not exercised. Long-term memory, also called secondary or reference memory, stores information for longer periods. The capacity of long-term memory is unlimited, and it can endure indefinitely. In addition, psychologists distinguish episodic memory, a memory of specific events or episodes normally described by the verb remember, from semantic memory, a knowledge of facts normally said to be known rather than remembered.

Memory is probably stored over wide areas of the brain rather than in any single location. However, amnesia, a memory disorder, can occur because of localized bilateral lesions in the limbic system – notably the hippocampus on the medial side of the temporal lobe, some parts of the thalamus, and their connections. This probably implies that these structures, rather than actually constituting a memory store, are important in the development of memories and in their recall. Memory impairment resulting from damage in these areas is a disorder of long-term episodic memory and is predominantly an anterograde amnesia – that is, it typically affects the memory of events occurring after the illness or accident causing the amnesia more than it does memories of the past. Substantial retrograde amnesia (loss of the memory of events occurring before the onset of the injury) rarely, if ever, occurs without significant anterograde amnesia as a result of brain damage, although it may occur alone in psychiatric disorders.

Although amnesia is a disorder of long-term episodic memory and leaves short-term and semantic memory intact, both of the latter can be affected by brain damage. Some parietal lobe lesions may affect short-term memory without affecting long-term memory. Short-term memory impairment – at least for verbal material – may be further subdivided into auditory and visual domains; however, these disorders result in difficulty in understanding spoken and written language rather than in memory impairment (i.e. they appear more like aphasia and dyslexia). Impairment of semantic memory also results in an impairment that resembles a loss of concepts or a language deficit more than it resembles a memory impairment. Some forms of visual agnosia have been interpreted as semantic memory impairment, because patients are unable to recognize objects such as chairs because they no longer "know" what chairs are or what they look like (or can no longer access that knowledge).

Infants make robust advances in both recognition memory and recall memory during their first year. In recognition memory, an infant is able to recognize a particular object they have seen a short time earlier (and hence will look at a new object rather than the older one if both are present side by side). Although newborns cannot remember objects seen more than a minute or two previously, their memory improves fairly rapidly over the first four or five months of life. By one month they are capable of remembering an object they saw 24 hours earlier, and by one year they can recognize an object they saw several days earlier. Three-month-old infants can remember an instrumental response, such as kicking the foot to produce a swinging motion in a toy, that they learned two weeks earlier, but they respond more readily if their memory is strengthened by repeated performances of the action.

By contrast, recall memory involves remembering (retrieving the representation, or mental image) an event or object that is not currently present. A major advance in recall memory occurs between the 8th and 12th months and underlies the child's acquisition of what Piaget called "the idea of the permanent object". This advance becomes apparent when an infant watches an adult hide an object under a cloth and must wait a short period of time before being allowed to reach for it. A six-month-old will not reach under the cloth for the hidden object, presumably because they have forgotten that the object was placed there. A one-year-old, however, will reach for the object even after a 30-second delay, presumably because they are able to remember its being hidden in the first place. These improvements in recall memory arise from the maturation of circuits linking various parts of the brain together. The improvements enable the infant to relate an event in their environment to a similar event in the past. As a result, they begin to anticipate their mother's positive reaction when the two are in close face-to-face interaction, and they behave as if inviting her to respond. The infant may also develop new fears, such as those of objects, people, or situations with which they are unfamiliar, i.e. which they cannot relate to past experiences using recall memory.

Evidence for stages of learning comes from observations of learners over relatively extended series of trials (or comparatively long periods). The empirical data suggest that several alterations in memory function occur even during a single trial. The process that commits information to memory also seems to have several stages.

Most theorists attribute at least three stages to memory function: immediate, short term, and long term. Immediate memory seems to last little more than a second or so. For example, subjects may be asked to remember where specific objects are located within a complex array they have just seen. Their performance shows that considerable information is retained only briefly, rapidly fading unless it is given special attention.

Short-term memory lasts about 15–30 seconds, as after looking up a telephone number. A person makes the call, discovers they have forgotten the number (perhaps in the midst of dialling), and has to look it up again. Nevertheless, such short-term retention does make information available long enough to be rehearsed; if the learner repeats it to themselves, the number can be transferred to some sort of longer-term storage.

Thus, rehearsal seems to facilitate transfer of data from short-term to long-term memory. Once committed to long-term memory, the results of learning tend to endure but can be abruptly abolished when specific parts of the brain are injured or removed; they also are vulnerable to interference from other learning. Nevertheless, conditioned responses may undergo little or no forgetting over periods of months or years. And electrical stimulation of the surgically exposed brain while someone is awake can make them remember experiences long thought forgotten. Recall is reported to be similarly enhanced during hypnosis.

Memory abnormality refers to any of the disorders that affect the ability to remember. Disorders of memory must have been known to the ancients and are mentioned in several early medical texts, but it was not until the closing decades of the nineteenth century that serious attempts were made to analyse them or to seek their explanation in terms of brain disturbances. Of the early attempts, the most influential was that of a French psychologist, Théodule-Armand Ribot, who, in his *Diseases of Memory* (1881, English translation 1882), endeavoured to account for memory loss as a symptom of progressive brain disease by embracing principles describing the evolution of memory function in the individual, as offered by an English neurologist, John Hughlings Jackson. Ribot wrote:

> The progressive destruction of memory follows a logical order – a law. It advances progressively from the unstable to the stable. It begins with the most recent recollections, which, being lightly impressed upon the nervous elements, rarely repeated and consequently having no permanent associations, represent organization in its feeblest form. It ends with the sensorial, instinctive memory, which, having become a permanent and integral part of the organism, represents organization in its most highly developed stage.

The statement, amounting to Ribot's "law" of regression (or progressive destruction) of memory, enjoyed a considerable vogue and is not without contemporary influence. The notion has been applied with some success to phenomena as diverse as the breakdown of memory for language in a disorder called aphasia and the gradual return of memory after brain concussion. It also helped to strengthen the belief that the neural basis of memory undergoes progressive strengthening or consolidation as a function of time. Yet students of retrograde amnesia (loss of memory for relatively old events) agree that Ribot's principle admits of many exceptions. In recovery from concussion of the brain, for example, the most recent memories are not always the first to return. It has proved difficult, moreover, to disentangle the effects of passage of time from those of rehearsal or repetition on memory.

A Russian psychiatrist, Sergey Sergeyevich Korsakov (Korsakoff), may have been the first to recognize that amnesia need not necessarily be associated with dementia (or loss of the ability to reason), as Ribot and many others had supposed. Korsakov described severe but relatively specific amnesia for recent and current events among alcoholics who showed no obvious evidence of shortcomings in intelligence and judgement. This disturbance, called Korsakoff's syndrome, has been reported for a variety of brain disorders aside from alcoholism and appears to result from damage in a relatively localized part of the brain.

The neurological approach may be combined with evidence of psychopathology to enrich understanding of memory function. Thus, a French neurologist, Pierre Janet, described amnesia sufferers who were apparently very similar to those observed by Korsakov but who gave no evidence of underlying brain disease. Janet also studied people who had lost memory of extensive periods in the past, also without evidence of organic disorder. He was led to regard these amnesias as hysterical, explaining them in terms of dissociation: a selective loss of access to specific memory data that seem to hold some degree of emotional significance. In his experience, reconnection of dissociated memories could as a rule be brought about by suggestion while the sufferer was under hypnosis. Freud regarded hysterical amnesia as arising from a protective activity or defence mechanism against unpleasant recollections; he came to call this sort of forgetting repression, and he later invoked it to account for the typical inability of adults to recollect their earliest years (infantile amnesia). He held that all forms of psychogenic (not demonstrably organic) amnesia eventually could resolve after prolonged sessions of talking (psychotherapy) and that hypnosis was neither essential nor necessarily in the amnesiac's best interest. Nevertheless, hypnosis (sometimes induced with the aid of drugs) has been widely used in the treatment of hysterical amnesia, particularly in time of war when only limited time is available.

Amnesia

Amnesia can be defined as loss of memory occurring most often as a result of damage to the brain from trauma, stroke, Alzheimer's disease, alcohol and drug toxicity, or infection. Amnesia may be anterograde, in which events following the causative trauma or disease are forgotten, or retrograde, in which events preceding the causative event are forgotten.

The condition also may be traced to severe emotional shock, in which case personal memories (e.g. identity) are affected. Such amnesia seems to represent a psychological escape from or denial of memories that might cause anxiety. These memories are not actually lost, because they can generally be recovered through psychotherapy or after the amnesic state has ended.

Occasionally amnesia may last for weeks, months, or even years, during which time someone may begin an entirely new life. Such protracted reactions are called fugue states. When recovered, the person is usually able to remember events that occurred prior to onset, but events of the fugue period are forgotten. Post-hypnotic amnesia, the forgetting of most or all events that occur while under hypnosis in response to a suggestion by the hypnotist, has long been regarded as a sign of deep hypnosis.

The common difficulty of remembering childhood experiences is sometimes referred to as childhood amnesia.

Short-Term Memory Loss

The so-called short-term memory is typically intact among amnesia sufferers. Such victims usually can repeat a short phrase or a series of words or numbers from immediate memory as adequately as anyone of comparable age and intelligence. Such an amnesic person can retain the gist of a question or request long enough to respond appropriately, unless, of course, there is enough delay in performance or attention is diverted. Evidently the ability to register information is intact, if this means availability of data in short-term memory. Thus, experimental psychologists who favour a sharp distinction between short-term and long-term storage systems contend that the primary deficit in amnesia is an inability to transfer information from short-term to long-term storage.

Other Forms of Memory Loss

Defect of memory is one of the most frequently observed symptoms of impaired brain function. It may be transitory, as after an alcoholic bout or an epileptic seizure; or it may be enduring, as after severe head injury or in association with brain disease. When there is impaired ability to store memories of new experiences (up to total loss of memory for recent events) the defect is termed anterograde amnesia. Retrograde loss may progressively abate or shrink if recovery begins, or it may gradually enlarge in scope, as in cases of progressive brain disease. Minor grades of memory defect are not uncommon after effects of severe head injury or infections such as encephalitis; typically they are shown in forgetfulness about recent events, in slow and insecure learning of new skills, and sometimes in a degree of persistent amnesia for events preceding the illness.

Apparently first described in 1964, transient global amnesia consists of an abrupt loss of memory lasting from a few seconds to a few hours, without loss of consciousness or other evidence of impairment. The individual is virtually unable to store new experience, suffering permanent absence of memory for the period of the attack. There is also a retrograde loss that may initially extend up to years preceding the attack. This deficit shrinks rapidly in the course of recovery but leaves an enduring gap in memory that seldom exceeds the three-quarters of an hour before onset. Thus the person is left with a persisting memory gap only for what happened during the attack itself and in a short period immediately preceding. Such attacks may be recurrent, are thought to result from transient reduction in blood supply in specific brain regions, and sometimes presage a stroke.

On recovery of consciousness after trauma, someone who has been knocked out by a blow on the head at first typically is dazed, confused, and imperfectly aware of their whereabouts and circumstances. This so-called post-traumatic confusional state may last from an hour or so up to several days or even weeks. While in this condition, the individual appears unable to store new memories; on recovery they commonly report total amnesia for the period of altered consciousness (post-traumatic amnesia). They are also apt to show retrograde amnesia that may extend over brief or quite long periods into the past, the duration seeming to depend on such factors as severity of injury and the sufferer's age. In the gradual course of recovery, memories are often reported to return in strict chronological sequence from the most remote to the most recent, as in Ribot's law. Yet this is by no means always the case; memories often seem to return haphazardly and to become gradually interrelated in the appropriate time sequence. The amnesia that remains seldom involves more than the events that occurred shortly before the accident, although in severe cases careful inquiry may reveal some residual memory defect for experiences dating from as long as a year before the trauma. It is thought by some that, after recovery, the overall period of time for which there is no recollection may indicate the degree of severity of the head injury.

Electroconvulsive treatments have been widely used in psychiatry, particularly for depressed people. A seizure or convulsion is induced by passing current through electrodes placed on the forehead. Each treatment is followed by a period of confusion for which the person is subsequently amnesic; at this time there is also a rapidly abating amnesia of some seconds for events that immediately preceded the shock. After a number of treatments, however, some individuals complain of more persistent memory defect, shown mainly in exaggerated forgetfulness for day-to-day events. These difficulties nearly always clear up within a few weeks after treatment ends. Experimental evidence tentatively suggests that electroshock administered to only one side of the head produces therapeutic results equal to those of the standard procedure but with significantly reduced impairment of memory.

First described in cases of chronic alcoholism, Korsakoff's psychosis, or syndrome, occurs in a wide variety of toxic and infectious brain illnesses, as well as in association with such nutritional disorders as deficiency of the B vitamins. The syndrome has also been observed among people with cerebral tumours, especially those involving the third ventricle (one of the fluid-filled cavities in the brain). The main psychological feature is gross defect in recent memory, sometimes so severe as to produce "moment-to-moment" consciousness; such people can store new information only for a few seconds and report no continuity between one experience and the next. They seem incapable of learning, even after many trials or repetitions. Although cases of such severity are relatively rare, the ability to store experience only briefly is quite characteristic of Korsakoff's syndrome.

Attention repeatedly has been drawn to severe and persistent memory defect following attacks of a form of brain inflammation called acute inclusion body encephalitis. The individual's behaviour closely resembles that of Korsakoff's syndrome except that their insight into the memory disorder is usually good and confabulation (confusion of real memories with imagined ones) is infrequent or absent. Indeed, the memory disorder is sometimes so limited and specific as to raise the possibility of a psychogenic (i.e. hysterical) amnesia. In cases of this kind there may be little or no impairment of intelligence or judgement.

Surgical operations on the sides of the brain (the temporal lobes) to remove tissues that produce symptoms of epilepsy are routine. While good results are often achieved, a degree of memory defect ensues. Operations on the dominant (usually left) temporal lobe tend to hamper one's ability to learn verbal information by hearing or reading. Usually observable even before surgery, the defect tends to be more marked after operation and has been reported to persist for up to three years before eventual recovery. Operations on one temporal lobe when there is unsuspected damage to its matching area on the other side of the brain (or on both lobes, in surgery very rarely undertaken) produce severe and persistent general memory defect, altogether comparable to post-encephalitic amnesia. There is gross defect in recent memory and in learning (except perhaps in motor learning), with retrograde amnesia that initially may involve several years of someone's past. Intelligence otherwise appears to be well preserved; the individual shows insight into their memory difficulty, and seldom, if ever, confabulates.

Déjà Vu

The déjà vu experience has aroused considerable interest and is occasionally felt by most people, especially in youth or when they are fatigued. It has also found its way into literature, having been well described by, among other creative writers, Shelley, Dickens, Hawthorne, Tolstoy, and Proust. The curious sense of extreme familiarity may be limited to a single sensory system, such as the sense of hearing, but as a rule it is generalized, affecting all aspects of experience including the subject's own actions. As a rule, it passes within a few seconds or minutes, though its repercussions may persist for some time. For some epileptics, however, déjà vu may continue for hours or even days and can provide a fertile subsoil for delusional elaboration.

In view of its occurrence among organically healthy individuals, déjà vu has commonly been regarded as psychogenic and as having its origin in some partly forgotten memory, fantasy, or dream. This explanation has appealed strongly to psychoanalysts; it also gains support from the finding that an experience very similar to déjà vu can be induced in normal people by hypnosis. If someone who has been hypnotized is presented with a picture and then told to forget it, but is then shown this and other pictures when they are awake, the person may report an intense feeling of familiarity that they are at a loss to justify. The déjà vu phenomenon also is attributable to minor neurophysiological abnormality; it is frequent in epilepsy. Indeed, déjà vu is accepted as a definite sign of epileptic activity originating in the temporal lobe of the brain and may occur as part of the seizure activity or frequently between convulsions. It seems to be more frequent in cases in which the disorder is in the right temporal lobe and has on occasion been evoked by electrical stimulation of the exposed brain during surgery. Some have been tempted to ascribe it to a dysrhythmic electrical discharge in some region of the temporal lobe that is closely associated with memory function.

Language

One of the most prominent intellectual activities dependent on memory is language. Although the neural basis of language is not fully understood, scientists have learned a great deal about this function of the brain from studies of patients who have lost speech and language abilities due to stroke, and from behavioural and functional neuroimaging studies of normal people.

A prominent and influential model, based on studies of these patients, proposes that (as described in Chapter 4), the underlying structure of speech comprehension arises in part of the left hemisphere of the brain called Wernicke's area. This temporal lobe region is connected with another region, Broca's area, in the frontal lobe, where a programme for vocal expression is created. Thus the language area of the brain surrounds the Sylvian fissure in the dominant hemisphere and is divided into these two major components named after Paul Broca and Carl Wernicke. The programme for vocal expression is then transmitted to a nearby area of the motor cortex that activates the mouth, tongue, and larynx.

This same model proposes that, when reading a word, the information is transmitted from the primary visual cortex to the angular gyrus, where the message is somehow matched with the words when they are spoken. The auditory form of the word is then processed for comprehension in Wernicke's area as if the word had been heard. Writing in response to an oral instruction requires information to be passed along the same pathways in the opposite direction – from the auditory cortex to Wernicke's area to the angular gyrus (see Figure 2.2). This model accounts for much of the data from patients and is the most widely used for clinical diagnosis and prognosis. Some refinements to this model may be necessary, however, because of both recent studies with patients and functional neuroimaging studies in healthy people.

Broca's area lies in the third frontal convolution, just anterior to the face area of the motor cortex and just above the Sylvian fissure. This is often described as the motor, or expressive, speech area; damage to it results in Broca aphasia, a language disorder characterized by deliberate, telegraphic speech with very simple grammatical structure, though the speaker may be quite clear as to what they wish to say and may communicate successfully. The Wernicke area is in the superior part of the posterior temporal lobe; it is close to the auditory cortex and is considered to be the receptive language, or language-comprehension, centre. An individual with Wernicke aphasia has difficulty understanding language; speech is typically fluent but is empty of content and characterized by circumlocutions, a high incidence of vague words like "thing", and sometimes neologisms and senseless "word salad". The entire posterior language area extends into the parietal lobe and is connected to Broca's area by a fibre tract called the arcuate fasciculus. Damage to this tract may result in conduction aphasia, a disorder in which the individual can understand and speak but has difficulty in repeating what is said to them. The suggestion is that, in this condition, language can be comprehended by the posterior zone and spoken by the anterior zone, but is not easily shuttled from one to the other.

Aphasia is a disorder of language and not of speech (although an apraxia of speech, in which the programming of motor speech output is affected, may accompany aphasia). The writing and reading of aphasic individuals, therefore, usually commit the same type of error as their speech, while the reverse is not the case. Isolated disorders of writing (dysgraphia) or, more commonly, reading (dyslexia) may occur as well, but these reflect a disruption of additional processing required for these activities over and above that required for language.

One particular form of dyslexia, dyslexia without dysgraphia, is an example of a disconnection syndrome – a disorder resulting from the disconnection of two areas of the brain rather than from damage to a centre. This type of dyslexia, also called letter-by-letter reading, is not associated with a writing disturbance; individuals tend to attempt to read by spelling words out loud, letter by letter. It usually results from a lesion in the posterior part of the left hemisphere that disconnects the visual areas of the brain from the language areas. This renders the language areas effectively blind, so that they cannot interpret visible language such as the written word. Writing is unaffected because the right hand is still connected to the left hemisphere, and, if letters can be spoken out loud correctly (which is not always the case), the individual will be able to hear themselves say them and reintegrate them into words. Disconnection syndromes are an important concept in understanding behavioural disorders associated with brain damage. The possibility that deficits are caused by disconnection must always be borne in mind.

Psycholinguistics

The term psycholinguistics was coined in the 1940s and came into more general use after the publication of Charles E. Osgood and Thomas A. Sebeok's *Psycholinguistics: A Survey of Theory and Research Problems* (1954), which reported the proceedings of a seminar sponsored in the USA by the Social Science Research Council's Committee on Linguistics and Psychology.

The boundary between linguistics and psycholinguistics is difficult, perhaps impossible, to draw. So too is the boundary between psycholinguistics and psychology. What characterizes psycholinguistics as it is practised today as a more or less distinguishable field of research is its concentration upon a certain set of topics connected with language and its bringing to bear upon them the findings and theoretical principles of both linguistics and psychology. The range of topics that would be generally held to fall within the field of psycholinguistics nowadays is rather narrower, however, than that covered in the survey by Osgood and Sebeok.

Language Acquisition by Children

One of the topics most central to psycholinguistic research is the acquisition of language by children. The term acquisition is preferred to "learning", because "learning" tends to be used by psychologists in a narrowly technical sense, and many psycholinguists believe that no psychological theory of learning, as currently formulated, is capable of accounting for the process whereby children, in a relatively short time, come to achieve a fluent control of their native language. Since the beginning of the 1960s, research on language acquisition has been strongly influenced by Chomsky's theory of generative grammar, and the main problem to which it has addressed itself has been how it is possible for young children to infer the grammatical rules underlying the speech they hear and then to use these rules for the construction of utterances that they have never heard before. It is Chomsky's conviction, shared by a number of psycholinguists, that children are born with a knowledge of the formal principles that determine the grammatical structure of all languages, and that it is this innate knowledge that explains the success and speed of language acquisition. Others have argued that it is not grammatical competence as such that is innate but more general cognitive principles and that the application of these to language utterances in particular situations ultimately yields grammatical competence. Many recent works have stressed that all children go through the same stages of language development regardless of the language they are acquiring. It has also been asserted that the same basic semantic categories and grammatical functions can be found in the earliest speech of children in a number of different languages operating in quite different cultures in various parts of the world.

Although Chomsky was careful to stress in his earliest writings that generative grammar does not provide a model for the production or reception of language utterances, there has been a good deal of psycholinguistic research directed toward validating the psychological reality of the units and processes postulated by generative grammarians in their descriptions of languages. Experimental work in the early 1960s appeared to show that non-kernel sentences took longer to process than kernel sentences and, even more interestingly, that the processing time increased proportionately with the number of optional transformations involved. More recent work has cast doubt on these findings, and most psycholinguists are now more cautious about using grammars produced by linguists as models of language processing. Nevertheless, generative grammar continues to be a valuable source of psycholinguistic experimentation, and the formal properties of language, discovered or more adequately discussed by generative grammarians than they have been by others, are generally recognized to have important implications for the investigation of short-term and long-term memory and perceptual strategies.

Speech Perception

Another important area of psycholinguistic research that has been strongly influenced by theoretical advances in linguistics is speech perception. It has long been realized that the identification of speech sounds and of the word forms composed of them depends upon the context in which they occur and upon the hearer's having mastered, usually as a child, the appropriate phonological and grammatical system. Throughout the 1950s, work on speech perception was dominated (as was psycholinguistics in general) by information theory, according to which the occurrence of each sound in a word and each word in an utterance is statistically determined by the preceding sounds and words. Information theory is no longer generally accepted, and subsequent research showed that in speech perception the cues provided by the acoustic input are interpreted, unconsciously and very rapidly, with reference not only to the phonological structure of the language but also to the more abstract levels of grammatical organization.

Other areas of psycholinguistics that should be briefly mentioned are the study of aphasia and neurolinguistics. The term aphasia is used to refer to various kinds of language disorders; recent work has sought to relate these, on the one hand, to particular kinds of brain injury and, on the other, to psychological theories of the storage and processing of different kinds of linguistic information. One linguist has put forward the theory that the most basic distinctions in language are those that are acquired first by children and are subsequently most resistant to disruption and loss in aphasia. This, though not disproved, is still regarded as controversial. Two kinds of aphasia are commonly distinguished. In motor aphasia the patient manifests difficulty in the articulation of speech or in writing and may produce utterances with a simplified grammatical structure, but their comprehension is not affected. In sensory aphasia the patient's fluency may be unaffected, but their comprehension will be impaired and their speech often incoherent.

Neurolinguistics

Neurolinguistics should perhaps be regarded as an independent field of research rather than as part of psycholinguistics. In 1864 it was shown that motor aphasia is produced by lesions in the third frontal convolution of the left hemisphere of the brain. Shortly after the connection had been established between motor aphasia and damage to this area (Broca's area), the source of sensory aphasia was localized in lesions of the posterior part of the left temporal lobe. More recent work has confirmed these findings. The technique of electrically stimulating the cortex in conscious patients has enabled brain surgeons to induce temporary aphasia and so to identify a "speech area" in the brain. It is no longer generally believed that there are highly specialized "centres" within the speech area, each with its own particular function; but the existence of such a speech area in the dominant hemisphere of the brain (which for most people is the left hemisphere) seems to be well established. The posterior part of this area is involved more in the comprehension of speech and the construction of grammatically and semantically coherent utterances, and the anterior part is concerned with the articulation of speech and with writing. Little is yet known about the operation of the neurological mechanisms underlying the storage and processing of language.

Speech Disorders

In accordance with physiological considerations, disorders of communication are first classified into disorders of voice and phonic respiration, disorders of articulated speech, and disorders of language. It has been known for a long time that the majority of communication disorders are not caused by local lesions of the teeth, tongue, vocal cords, or regulating brain centres. Because these predominant disorders of voice and speech develop from derangements of the underlying physiological functions of breathing, use of the voice, speaking habits, or emotional disorders, this group has been labelled as functional. The remainder of the communication disorders with clearly recognizable structural abnormalities in the total speech mechanism has been termed organic.

While this empirical grouping has certain implications for the selection of the appropriate treatment, it is not satisfactory because organic structure and living function can never be separated. Certain functional disorders of the voice caused by its habitual abuse may very well lead to secondary structural changes, such as the growths (polyps and nodules) of the vocal cords, which develop as a result of vocal abuse. On the other hand, all of the obviously organic and structural lesions, such as loss of the tongue from accident or surgery, almost inevitably will be followed by emotional and other psychological reactions. In this case, the functional components are of secondary nature but to a great extent will influence the total picture of disturbance, including the patient's ability to adjust to their limitation, to relearn a new mode of appropriate function, and to make the best of their condition.

Within these major groups, the various types of communication disorders have for a long time, and in most parts of the world, been described by the listener's perceptual impression. Most languages employ specific words for the various types of abnormal speech, such as stuttering, stammering, cluttering, mumbling, lisping, whispering, and many others. The problem with such subjective and symptomatic labels is the fact that they try to define the final, audible result, the recognizable phenomenon, and not by any means the underlying basis. This general human tendency to describe disorders of communication by what the listener hears is analogous to the attempts of early medicine to classify diseases by the patient's symptoms that the diagnosing physician could see or hear or feel or perhaps smell. Before the great discoveries of the nineteenth century had erected a logical basis for medical pathology, the various diseases were classified as numerous types of fevers, congestions, dyscrasias, etc. Thus, malaria was originally thought to be caused by the evil emanations (miasma) of the bad air (*mal aria*) near swamps, until it was recognized to be caused by a blood parasite transmitted by the mosquito.

The various approaches of medical, psychiatric, psychological, educational, behavioural, and other schools of speech pathology have made great advances in the recent past and better systems of classification continue to be proposed. They aim at grouping the observable symptoms of speech disorders according to the underlying origins instead of the listener's subjective impressions. While this is relatively easy in the case of language loss from, for example, a brain stroke because the destroyed brain areas can be identified at autopsy, it is more difficult in the case of the large group of so-called functional speech disorders for two reasons: first, they are definitely not caused by gross, easily visible organic lesions, and second, many functional disorders are outgrown through maturation or appropriate learning (laboratory study of the involved tissues in such cases would reveal no detectable lesions). It is hoped that refined methods of study in the areas of both "functional" psychology and "organic" neurophysiology will eventually reveal the structural bases for the prevalent disorders of voice and speech.

The most frequent speech disorders are those that disturb the child's acquisition or learning of language. While those concerned with modern terminology are striving for an improved classification according to the aetiological (causative) factors, it is still customary to classify these disorders on the basis of the complaint: absence of speech, baby talk, poorly intelligible articulation, lisping, etc. Recent studies of large numbers of children with such developmental language disorders have shown that at least two chief classes of these disorders may be distinguished: general language disability from genetic factors with a familial (inherited) pattern chiefly from the paternal side, and acquired language disorders due to damage sustained before, during, or shortly after birth (i.e. perinatally).

These latter perinatal damages encompass the gamut of toxic, infectious, traumatic, nutritional, hormonal, and other damages that may hurt the growing fetus or young infant. Major and minor birth injury is not an infrequent factor. Hereditary factors also encompass a great variety of genetically predetermined influences, including familial tendency to exhibit slow language development, lesser endowment in the brain area for language, inferior function in the highest brain areas of auditory performance without organic damage to the ears, slow maturation of motor function (including clumsiness and deviation from normal cerebral dominance), and other signs of delayed cerebral growth. Additional environmental causes include poor language patterns used by the family, parental neglect, emotional maladjustment, general weakness from prolonged disease, as well as various socioeconomic, cultural, and other psychological influences.

While some otherwise perfectly normal children, particularly boys, may not elect to begin talking until age three, making good progress in every respect from then on, the absence of speech after age two may be caused by any of the conditions mentioned thus far and would appear to merit prompt investigation. If an organic cause can be detected, the symptomatic description of delayed language development then yields to a specific aetiological (causal) diagnosis. Although it is best to describe the absence of speech in early childhood as simply delayed language development, some American investigators tend to refer to this condition as congenital (present at birth) aphasia, a term rejected by most European scholars who argue that there cannot be an inborn or early acquired aphasia before a language has been learned.

Many children encounter unusual difficulties in mastering the patterns of articulation of their mother tongue and are said to manifest articulatory immaturity (infantile dyslalia). If no organic cause can be found, the probable cause may be delayed maturation of psychomotor skills.

Marked delays of language development are often followed by a period of inability to learn the rules of grammar and syntax at the usual age (dysgrammatism). Though this is often a sign of inherited language disability, it may reflect intellectual disability or other types of brain damage.

Some children who have suffered such laboured language development may then go through a period of retarded reading and writing disability, a condition often defined as dyslexia. Again, there are two chief varieties: the primary or developmental reading and writing disability due to constitutional (organic) and hereditary factors, and a large secondary group of symptomatic reading disorders acquired through any of the influences that retard language development in general, including troubles with vision. Practically all investigators agree that primary or developmental dyslexia shows a marked hereditary tendency (is familial) and is typically associated with other disorders of psychomotor development, deviation from the prevalent right-handedness, and poor function in the auditory area in the brain, often associated with lack of musical talent. Primary dyslexia is significantly associated with other developmental speech disorders.

The treatment of stuttering is difficult and often demands much skill and responsibility on the part of the therapist. The possibility of some specific medical cure seems remote at the present time. Even the most advanced methods of modern psychiatry have failed to produce superior results in treatment. For a time it was hoped that new psychopharmacological drugs (e.g. tranquillizers) might facilitate and accelerate recovery from stuttering, but these efforts have been disappointing thus far. The typical programme of management in this disorder is a strict one, of psychotherapy (talking freely with a psychiatrist or psychologist so as to reduce emotional problems) supported by various applications of learning theory or behavioural theory (in retraining the stutterer) and other techniques depending on the therapist's position. It is widely agreed that the patient must acquire a better adjustment to the problems of their life and that they need to develop a technique for controlling their symptoms and fears. Prognosis (predicted outcome of treatment) thus is held to depend greatly on the patient's motivation and perseverance. It is interesting to note that experienced investigators no longer aspire to a "cure" of stuttering through an aetiological (causal) approach. Instead of focusing on underlying causes, they aim at making the patient "symptom-free" via symptomatic therapy.

Dysphasia means the partial or total loss of language as a result of lesions in those parts of the brain that are directly related to language function. Stroke in elderly patients and head injury in younger ones are typical causes. Aphasia is seen most frequently when the left side of the brain is afflicted, as evidenced by paralysis of the right arm and leg. Evidence indicates that the left hemisphere is dominant in all right-handed individuals and in some left-handers as well. Some experts even believe that the left brain hemisphere is dominant for language in most individuals regardless of handedness and that dominance of the right brain is exceptional in some left-handers. According to other opinions, dominance for language is more evenly distributed in both hemispheres in left-handed people. As explained before, two major brain areas are recognized as intimately associated with language function: Broca's area in the third frontal convolution and Wernicke's area in the posterior third of the upper temporal convolution. The angular gyrus at the junction of the temporal (side), parietal (top), and occipital (back) lobes of the brain is believed to be related to graphic language as used for reading and writing.

Aphasiology, the science of aphasic language loss, is studied by neurologists, neurosurgeons, some phoniatrists, certain speech pathologists, as well as some psychologists and linguists. This diversity of research background accounts, in part, for the great diversity in theoretical approaches to aphasia. Numerous classification schemes that have been proposed vary from simple groupings into a few main types of aphasia to complicated systems with many forms and subtypes of aphasic disturbances. Similar to research in stuttering, the literature on aphasia is exceptionally large and growing.

The essence of aphasia is the loss of memory for the meaning of language and its production. Thus, in the predominantly expressive or motor forms, the patient can no longer remember the intricate patterns for articulation; they can no longer form a word in speaking or writing, even though they may know what they want to express. In the predominantly receptive or sensory forms, the patient can talk freely, sometimes excessively and incessantly (logorrhea), although with numerous errors and meaningless clichés, but they no longer comprehend what is said to them or what they try to read. Those who recover from receptive forms of the disorder are likely to explain that during their aphasia spoken language sounded like an unintelligible, alien tongue. The degree to which there is combination of expressive and receptive symptoms varies greatly with the type and extent of brain lesion. There may be total loss of all language functions (global aphasia) to slight residual errors or misunderstandings when the brain damage is only slight or temporary. A major complication of aphasia is the frequent association with right hemiplegia, in which the paralysed hand is no longer serviceable for writing. Retraining of the left hand for writing may then become necessary.

Management of aphasia has two goals: one, the physical recovery of the patient through treatment by the internist, neurologist, and possibly brain surgeon, and two, the re-education of the brain functions that are still present, the aim of which is to help the patient relearn some use of language under the guidance of the speech pathologist. The better the patient's recovery from the brain lesion, the more chances there are of prompt and complete return of language. When the brain losses are permanent, the patient must relearn each word, sentence, and phrase like the young child, albeit in a more cumbersome manner, apparently using parts of the brain that still function. They are repeatedly shown the picture of an object along with its printed name; these words are spoken to them by their teacher over and over in the hope that the patient will learn to repeat the word or phrase until they can say it spontaneously. Words that are relevant to the patient's every-day life are emphasized first with due consideration for their interests and past occupation; later they may advance to the use of abstract concepts and of higher levels of language. Various types of automated training programmes are avail-able, including a simple play-back device that shows an inserted card with a picture on it, the name of the pictured object printed next to it, and the audible word recorded on a strip of magnetic tape. Devices of this type enable the patient to practise at their own pace.

Under this heading may be summarized various types of communication disorders that develop on the basis of known structural lesions or metabolic disturbances. Aetiological clas-sifications group these impediments according to the types of organic diseases, as well as in respect to the afflicted effector organs (such as the tongue). Disturbed speech from lesions in the various parts of the nervous system is known as dysarthria. Intellectual disability usually limits the development of linguis-tic ability to the same extent as it does intellectual capacity; this language disorder has been described as dyslogia. Mental disturbances can also manifest themselves in linguistic symptoms, such as in the peculiar (dysphrenic) mode of speech among sufferers of schizophrenia. Hearing loss dating from early childhood leads to a typical distortion of the speech pattern for which various names have been coined, such as audiogenic dyslalia. Visible defects in such oral articulators as the lips and teeth limit the mechanics of articulation and thus reduce the quality and intelligibility of speech; such speech problems are known collectively as dysglossia.

Damage to those parts of the nervous system that regulate the actions of voice and speech cause distinctive alterations of the speech pattern. The most important disorder of this type is cerebral palsy from brain injury before, during, or soon after birth. The majority of cerebral palsy victims retain normal intelligence but are handicapped by distortions of voluntary movements, including those for speaking. Just as walking may be stilted and jerky and arm movements crude and uncontrolled, the patterns of voice and speech will reflect the same distortions. Great advances in rehabilitation have been achieved in the recent past, such as with the well-known Bobath method, which is based on learned suppression of primitive reflexes.

Expert analysis of the bizarre speech patterns associated with certain psychiatric disturbances is of primary diagnostic significance. If a mute child persists in stereotyped rituals and strange behaviour, a diagnosis of childhood autism is likely to be made. Some experts distinguish this from a similar disorder called childhood schizophrenia, in which previously good general and linguistic development falls apart in association with similarly bizarre behaviour. In adolescence, a sudden change of voice to a shrill falsetto or weird chanting may herald the outbreak of juvenile schizophrenic disease. Infantile lisping, strange distortions of articulation, and various eccentricities in verbal expression are other signs of schizophrasic speech in the adult schizophrenic. Linguistic study may help the psychiatrist analyse the patient's ways of thinking and to provide a measure of the sufferer's progress under therapy. Great therapeutic stress is placed on establishing contact with the autistic child by eliciting from them some sort of communication. Language is felt to be one of the best bridges to break open the closed inner world of such children.

7

INTELLIGENCE –
HUMAN AND ARTIFICIAL

Human intelligence is the mental quality that consists of the abilities to learn from experience, adapt to new situations, understand and handle abstract concepts, and use knowledge to manipulate one's environment.

Much of the excitement among investigators in the field of intelligence derives from their attempts to determine exactly what intelligence is. Different investigators have emphasized different aspects of intelligence in their definitions. For example, in a 1921 symposium the American psychologists Lewis M. Terman and Edward L. Thorndike differed over the definition of intelligence, Terman stressing the ability to think abstractly and Thorndike emphasizing learning and the ability to give good responses to questions.

More recently, however, psychologists have generally agreed that adaptation to the environment is the key to understanding both what intelligence is and what it does. Such adaptation may occur in a variety of settings: a student in school learns the material they need to know in order to do well in a course; a physician treating a patient with unfamiliar symptoms learns about the underlying disease; or an artist reworks a painting to convey a more coherent impression. For the most part, adaptation involves making a change in oneself in order to cope more effectively with the environment, but it can also mean changing the environment or finding an entirely new one.

Effective adaptation draws upon a number of cognitive processes, such as perception, learning, memory, reasoning, and problem solving. The main emphasis in a definition of intelligence, then, is that it is not a cognitive or mental process per se but rather a selective combination of these processes that is purposively directed toward effective adaptation. Thus, the physician who learns about a new disease adapts by perceiving material on the disease in medical literature, learning what the material contains, remembering the crucial aspects that are needed to treat the patient, and then utilizing reason to solve the problem of applying the information to the needs of the patient. Intelligence, in total, has come to be regarded not as a single ability but as an effective drawing together of many abilities. This has not always been obvious to investigators of the subject, however; indeed, much of the history of the field revolves around arguments regarding the nature and abilities that constitute intelligence.

Theories of Intelligence

Theories of intelligence, as is the case with most scientific theories, have evolved through a succession of models. Four of the most influential paradigms have been psychological measurement, also known as psychometrics; cognitive psychology, which concerns itself with the processes by which the mind functions; cognitivism and contextualism, a combined approach that studies the interaction between the environment and mental processes; and biological science, which considers the neural bases of intelligence. What follows is a discussion of developments within these four areas.

Psychometric Theories

Psychometric theories have generally sought to understand the structure of intelligence: What form does it take, and what are its parts, if any? Such theories have generally been based on and established by data obtained from tests of mental abilities, including analogies (e.g. lawyer is to client as doctor is to __), classifications (e.g. Which word does not belong with the others? robin, sparrow, chicken, bluejay), and series completions (e.g. What number comes next in the following series? 3, 6, 10, 15, 21,_).

Psychometric theories are based on a model that portrays intelligence as a composite of abilities measured by mental tests. This model can be quantified. For example, performance on a number-series test might represent a weighted composite of number, reasoning, and memory abilities for a complex series. Mathematical models allow for weakness in one area to be offset by strong ability in another area of test performance. In this way, superior ability in reasoning can compensate for a deficiency in number ability.

One of the earliest of the psychometric theories came from the British psychologist Charles E. Spearman (1863–1945), who published his first major article on intelligence in 1904. He noticed what may seem obvious now – that people who did well on one mental ability test tended to do well on others, while people who performed poorly on one of them also tended to perform poorly on others. To identify the underlying sources of these performance differences, Spearman devised factor analysis, a statistical technique that examines patterns of individual differences in test scores. He concluded that just two kinds of factors underlie all individual differences in test scores. The first and more important factor, which he labelled the "general factor", or g, pervades performance on all tasks requiring intelligence. In other words, regardless of the task, if it requires intelligence, it requires g. The second factor is specifically related to each particular test. For example, when someone takes a test of arithmetical reasoning, their performance on the test requires a general factor that is common to all tests (g) and a specific factor that is related to whatever mental operations are required for mathematical reasoning as distinct from other kinds of thinking. But what, exactly, is g? After all, giving something a name is not the same as understanding what it is. Spearman did not know exactly what the general factor was, but he proposed in 1927 that it might be something like "mental energy".

The American psychologist L.L. Thurstone disagreed with Spearman's theory, arguing instead that there were seven factors, which he identified as the "primary mental abilities". These seven abilities, according to Thurstone, were verbal comprehension (as involved in the knowledge of vocabulary and in reading), verbal fluency (as involved in writing and in producing words), number (as involved in solving fairly simple numerical computation and arithmetical reasoning problems), spatial visualization (as involved in visualizing and manipulating objects, such as fitting a set of suitcases into an automobile trunk), inductive reasoning (as involved in completing a number series or in predicting the future on the basis of past experience), memory (as involved in recalling people's names or faces), and perceptual speed (as involved in rapid proofreading to discover typographical errors in a text).

Although the debate between Spearman and Thurstone has remained unresolved, other psychologists – such as Canadian Philip E. Vernon and American Raymond B. Cattell – have suggested that both were right in some respects. Vernon and Cattell viewed intellectual abilities as hierarchical, with g, or general ability, located at the top of the hierarchy. But below g are levels of gradually narrowing abilities, ending with the specific abilities identified by Spearman. Cattell, for example, suggested in *Abilities: Their Structure, Growth, and Action* (1971) that general ability can be subdivided into two further kinds, "fluid" and "crystallized". Fluid abilities are the reasoning and problem-solving abilities measured by tests such as analogies, classifications, and series completions. Crystallized abilities, which are thought to derive from fluid abilities, include vocabulary, general information, and knowledge about specific fields. The American psychologist John L. Horn suggested that crystallized abilities more or less increase over a person's lifespan, whereas fluid abilities increase in earlier years and decrease in later ones.

Most psychologists agreed that Spearman's subdivision of abilities was too narrow, but not all agreed that the subdivision should be hierarchical. The American psychologist Joy Paul Guilford proposed a structure-of-intellect theory, which in its earlier versions postulated 120 abilities. In *The Nature of Human Intelligence* (1967), Guilford argued that abilities can be divided into five kinds of operation, four kinds of content, and six kinds of product. These facets can be variously combined to form 120 separate abilities. An example of such an ability would be cognition (operation) of semantic (content) relations (product), which would be involved in recognizing the relation between lawyer and client in the analogy problem above (lawyer is to client as doctor is to __). Guilford later increased the number of abilities proposed by his theory to 150.

Eventually it became apparent that there were serious problems with the basic approach to psychometric theory. A movement that had started by postulating one important ability had come, in one of its major manifestations, to recognize 150. Moreover, the psychometricians (as practitioners of factor analysis were called) lacked a scientific means of resolving their differences. Any method that could support so many theories seemed somewhat suspect. Most important, however, the psychometric theories failed to say anything substantive about the processes underlying intelligence. It is one thing to discuss "general ability" or "fluid ability" but quite another to describe just what is happening in people's minds when they are exercising the ability in question. The solution to these problems, as proposed by cognitive psychologists, was to study directly the mental processes underlying intelligence and, perhaps, to relate them to the facets of intelligence posited by psychometricians.

The American psychologist John B. Carroll, in *Human Cognitive Abilities* (1993), proposed a "three-stratum" psychometric model of intelligence that expanded upon existing theories of intelligence. Many psychologists regard Carroll's model as definitive, because it is based upon re-analyses of hundreds of data sets. In the first stratum, Carroll identified narrow abilities (roughly 50 in number) that included the seven primary abilities identified by Thurstone. According to Carroll, the middle stratum encompassed broad abilities (approximately 10) such as learning, retrieval ability, speediness, visual perception, fluid intelligence, and the production of ideas. The third stratum consisted solely of the general factor, g, as identified by Spearman. It might seem self-evident that the factor at the top would be the general factor, but it is not, because there is no guarantee that there is any general factor at all.

Both traditional and modern psychometric theories face certain problems. First, it has not been proved that a truly general ability encompassing all mental abilities actually exists. In *The General Factor of Intelligence: How General Is It?* (2002), edited by the psychologists Robert Sternberg and Elena Grigorenko, contributors to the edited volume provided competing views of the g factor, with many suggesting that specialized abilities are more important than a general ability, especially because they more readily explain individual variations in intellectual functioning. Second, psychometric theories cannot precisely characterize all that goes on in the mind. Third, it is not clear whether the tests on which psychometric theories are based are equally appropriate in all cultures. In fact, there is an assumption that successful performance on a test of intelligence or cognitive ability will depend on one's familiarity with the cultural framework of those who wrote the test. In her 1997 paper *You Can't Take It with You: Why Ability Assessments Don't Cross Cultures,* the American psychologist Patricia M. Greenfield concluded that a single test may measure different abilities in different cultures. Her findings emphasized the importance of taking issues of cultural generality into account when creating abilities tests.

Cognitive Theories

During the era dominated by psychometric theories, the study of intelligence was influenced most by those investigating individual differences in people's test scores. In an address to the American Psychological Association in 1957, the American researcher Lee Cronbach, a leader in the testing field, decried the lack of common ground between psychologists who studied individual differences and those who studied commonalities in human behaviour. Cronbach's plea to unite the "two disciplines of scientific psychology" led, in part, to the development of cognitive theories of intelligence and of the underlying processes posited by these theories.

Fair assessments of performance require an understanding of the processes underlying intelligence; otherwise, there is a risk of arriving at conclusions that are misleading, if not simply wrong, when evaluating overall test scores or other assessments of performance. Suppose, for example, that a student performs poorly on the verbal analogies questions in a psychometric test. One possible conclusion is that the student does not reason well, but an equally plausible interpretation is that the student does not understand the words or is unable to read them in the first place. A student who fails to solve the analogy "audacious is to pusillanimous as mitigate is to __" might be an excellent reasoner but have only a modest vocabulary, or vice versa. By using cognitive analysis, the test interpreter is able to determine the degree to which the poor score stems from low reasoning ability and the degree to which it results from not understanding the words.

Underlying most cognitive approaches to intelligence is the assumption that intelligence comprises mental representations (such as propositions or images) of information and processes that can operate on such representations. A more intelligent person is assumed to represent information more clearly and to operate faster on these representations. Researchers have sought to measure the speed of various types of thinking. Through mathematical modelling, they divide the overall time required to perform a task into the constituent times needed to execute each mental process. Usually, they assume that these processes are executed serially (one after another) and, hence, that the processing times are additive. But some investigators allow for parallel processing, in which more than one process is executed at the same time. Regardless of the type of model used, the fundamental unit of analysis is the same – that of a mental process acting upon a mental representation.

A number of cognitive theories of intelligence have been developed. Among them is that of the American psychologists Earl B. Hunt, Nancy Frost, and Clifford E. Lunneborg, who in 1973 showed one way in which psychometrics and cognitive modelling could be combined. Instead of starting with conventional psychometric tests, they began with tasks that experimental psychologists were using in their laboratories to study the basic phenomena of cognition, such as perception, learning, and memory. They showed that individual differences in these tasks, which had never before been taken seriously, were in fact related (although rather weakly) to patterns of individual differences in psychometric intelligence test scores. Their results suggested that the basic cognitive processes are the building blocks of intelligence.

The following example illustrates the kind of task Hunt and his colleagues studied in their research: the subject is shown a pair of letters, such as "A A", "A a", or "A b". The subject's task is to respond as quickly as possible to one of two questions: "Are the two letters the same physically?" or "Are the two letters the same only in name?" In the first pair the letters are the same physically, and in the second pair the letters are the same only in name.

The psychologists hypothesized that a critical ability underlying intelligence is the rapid retrieval of lexical information, such as letter names, from memory. Hence, they were interested in the time needed to react to the question about letter names. By subtracting the reaction time to the question about physical match from the reaction time to the question about name match, they were able to isolate and set aside the time required for sheer speed of reading letters and pushing buttons on a computer. They found that the score differences seemed to predict psychometric test scores, especially those on tests of verbal ability such as reading comprehension. Hunt, Frost, and Lunneborg concluded that verbally facile people are those who are able to absorb and then retrieve from memory large amounts of verbal information in short amounts of time. The time factor was the significant development in this research.

A few years later, Sternberg suggested an alternative approach that could resolve the weak relation between cognitive tasks and psychometric test scores. He argued that Hunt and his colleagues had tested for tasks that were limited to low-level cognitive processes. Although such processes may be involved in intelligence, Sternberg claimed that they were peripheral rather than central. He recommended that psychologists study the tasks found on intelligence tests and then identify the mental processes and strategies people use to perform those tasks.

Sternberg began his study with the analogies cited earlier: "lawyer is to client as doctor is to __". He determined that the solution to such analogies requires a set of component cognitive processes that he identified as follows: encoding of the analogy terms (e.g. retrieving from memory attributes of the terms lawyer, client, and so on); inferring the relation between the first two terms of the analogy (e.g. figuring out that a lawyer provides professional services to a client); mapping this relation to the second half of the analogy (e.g. figuring out that both a lawyer and a doctor provide professional services); applying this relation to generate a completion (e.g. realizing that someone to whom a doctor provides professional services is a patient); and then responding. By applying mathematical modelling techniques to reaction-time data, Sternberg isolated the components of information processing. He determined whether each experimental subject did, indeed, use these processes, how the processes were combined, how long each process took, and how susceptible each process was to error. Sternberg later showed that the same cognitive processes are involved in a wide variety of intellectual tasks. He subsequently concluded that these and other related processes underlie scores on intelligence tests.

A different approach was taken in the work of the British psychologist Ian Deary, among others. He argued that inspection time is a particularly useful means of measuring intelligence. It is thought that individual differences in intelligence may derive in part from differences in the rate of intake and processing of simple stimulus information. In the inspection-time task, a person looks at two vertical lines of unequal length and is asked to identify which of the two is longer. Inspection time is the length of time of stimulus presentation each individual needs in order to discriminate which of the two lines is the longest. Some research suggests that more intelligent individuals are able to discriminate the lengths of the lines in shorter inspection times.

Other cognitive psychologists have studied human intelligence by constructing computer models of human cognition. Two leaders in this field were the American computer scientists Allen Newell and Herbert A. Simon. In the late 1950s and early '60s, they worked with computer expert Cliff Shaw to construct a computer model of human problem solving. Called the General Problem Solver, it could find solutions to a wide range of fairly structured problems, such as logical proofs and mathematical word problems. This research, based on a heuristic procedure called "means–ends analysis", led Newell and Simon to propose a general theory of problem solving in 1972.

Most of the problems studied by Newell and Simon were fairly well structured, in that it was possible to identify a discrete set of steps that would lead from the beginning to the end of a problem. Other investigators have been concerned with other kinds of problems, such as how a text is comprehended or how people are reminded of things they already know when reading a text. The psychologists Marcel Just and Patricia Carpenter, for example, showed that complicated intelligence test items, such as figural matrix problems involving reasoning with geometric shapes, could be solved by a sophisticated computer program at a level of accuracy comparable to that of human test takers. It is in this way that a computer reflects a kind of "intelligence" similar to that of humans. One critical difference, however, is that programmers structure the problems for the computer, and they also write the code that enables the computer to solve the problems. Humans "encode" their own information and do not have personal programmers managing the process for them. To the extent that there is a "programmer", it is in fact the person's own brain.

All of the cognitive theories described so far rely on what psychologists call the "serial processing of information", meaning that in these examples, cognitive processes are executed in series, one after another. Yet the assumption that people process chunks of information one at a time may be incorrect. Many psychologists have suggested instead that cognitive processing is primarily parallel. It has proved difficult, however, to distinguish between serial and parallel models of information processing (just as it had been difficult earlier to distinguish between different factor models of human intelligence). Advanced techniques of mathematical and computer modelling were later applied to this problem. Possible solutions have included "parallel distributed processing" models of the mind, as proposed by the psychologists David E. Rumelhart and Jay L. McClelland. These models postulated that many types of information processing occur within the brain at once, rather than just one at a time.

Computer modelling has yet to resolve some major problems in understanding the nature of intelligence, however. For example, the American psychologist Michael E. Cole and other psychologists have argued that cognitive processing does not accommodate the possibility that descriptions of intelligence may differ from one culture to another and across cultural subgroups. Moreover, common experience has shown that conventional tests, even though they may predict academic performance, cannot reliably predict the way in which intelligence will be applied (i.e. through performance in jobs or other life situations beyond school). In recognition of the difference between real-life and academic performance, then, psychologists have come to study cognition not in isolation but in the context of the environment in which it operates.

Cognitive-Contextual Theories

Cognitive-contextual theories deal with the way that cognitive processes operate in various settings. Two of the major theories of this type are that of the American psychologist Howard Gardner and that of Sternberg. In 1983 Gardner challenged the assumption of a single intelligence by proposing a theory of "multiple intelligences". Earlier theorists had gone so far as to contend that intelligence comprises multiple abilities. But Gardner went one step further, arguing that intelligences are multiple and include, at a minimum, linguistic, logical-mathematical, spatial, musical, bodily-kinesthetic, interpersonal, and intrapersonal intelligence.

Some of the intelligences proposed by Gardner resembled the abilities proposed by psychometric theorists, but others did not. For example, the idea of a musical intelligence was relatively new, as was the idea of a bodily-kinesthetic intelligence, which encompassed the particular abilities of athletes and dancers. Gardner derived his set of intelligences chiefly from studies of cognitive processing, brain damage, exceptional individuals, and cognition across cultures. He also speculated on the possibility of an existential intelligence (a concern with "ultimate" issues, such as the meaning of life), although he was unable to isolate an area of the brain that was dedicated to the consideration of such questions. Gardner's research on multiple intelligences led him to claim that most concepts of intelligence had been ethnocentric and culturally biased but that his was universal, because it was based upon biological and cross-cultural data as well as upon data derived from the cognitive performance of a wide array of people.

An alternative approach that took similar account of cognition and cultural context was Sternberg's "triarchic" theory, which he proposed in *Beyond IQ: A Triarchic Theory of Human Intelligence* (1985). Both Gardner and Sternberg believed that conventional notions of intelligence were too narrow; Sternberg, however, questioned how far psychologists should go beyond traditional concepts, suggesting that musical and bodily-kinesthetic abilities are talents rather than intelligences because they are fairly specific and are not prerequisites for adaptation in most cultures.

Sternberg posited three ("triarchic") integrated and interdependent aspects of intelligence, which are concerned, respectively, with a person's internal world, the external world, and experience. The first aspect comprises the cognitive processes and representations that form the core of all thought. The second aspect consists of the application of these processes and representations to the external world. The triarchic theory holds that more intelligent people are not just those who can execute many cognitive processes quickly or well; rather, their greater intelligence is reflected in knowing their strengths and weaknesses and capitalizing upon their strengths while compensating for their weaknesses. More intelligent people, then, find a niche in which they can operate most efficiently. The third aspect of intelligence consists of the integration of the internal and external worlds through experience. This includes the ability to apply previously learned information to new or wholly unrelated situations.

Some psychologists believe that intelligence is reflected in an ability to cope with relatively novel situations. This explains why experience can be so important. For example, intelligence might be measured by placing people in an unfamiliar culture and assessing their ability to cope with the new situation. According to Sternberg, another facet of experience that is important in evaluating intelligence is the automatization of cognitive processing, which occurs when a relatively novel task becomes familiar. The more someone automatizes the tasks of daily life, the more mental resources they will have for coping with novelty.

Other intelligences were proposed in the late twentieth century. In 1990 the psychologists John Mayer and Peter Salovey defined the term emotional intelligence as the ability to perceive emotions, to access and generate emotions so as to assist thought, to understand emotions and emotional knowledge, and to reflectively regulate emotions so as to promote emotional and intellectual growth.

The four aspects identified by Mayer and Salovey involve (a) recognizing one's own emotions as well as the emotions of others, (b) applying emotion appropriately to facilitate reasoning, (c) understanding complex emotions and their influence on succeeding emotional states, and (d) having the ability to manage one's emotions as well as those of others. The concept of emotional intelligence was popularized by the psychologist and journalist Daniel Goleman in books published from the 1990s. Several tests developed to measure emotional intelligence have shown modest correlations between emotional intelligence and conventional intelligence.

Biological Theories

The theories discussed above seek to understand intelligence in terms of hypothetical mental constructs, whether they are factors, cognitive processes, or cognitive processes in interaction with context. Biological theories represent a radically different approach that dispenses with mental constructs altogether. Advocates of such theories, usually called reductionists, believe that a true understanding of intelligence is possible only by identifying its biological basis. Some would argue that there is no alternative to reductionism if, in fact, the goal is to explain rather than merely to describe behaviour. But the case is not an open-and-shut one, especially if intelligence is viewed as something more than the mere processing of information. As Howard Gardner pointedly asked in the article *What We Do and Don't Know About Learning* (2004):

> Can human learning and thinking be adequately reduced to the operations of neurons, on the one hand, or to chips of silicon, on the other? Or is something crucial missing, something that calls for an explanation at the level of the human organism?

Analogies that compare the human brain to a computer suggest that biological approaches to intelligence should be viewed as complementary to, rather than as replacing, other approaches. For example, when a person learns a new German vocabulary word, they become aware of a pairing, say, between the German term *Die Farbe* and the English word colour, but a trace is also laid down in the brain that can be accessed when the information is needed. Although relatively little is known about the biological bases of intelligence, progress has been made on three different fronts, all involving studies of brain operation.

Measuring Intelligence

Almost all of the theories discussed above employ complex tasks for gauging intelligence in both children and adults. Over time, theorists chose particular tasks for analysing human intelligence, some of which have been explicitly discussed here, e.g. recognition of analogies, classification of similar terms, extrapolation of number series, performance of transitive inferences, and the like. Although the kinds of complex tasks discussed so far belong to a single tradition for the measurement of intelligence, the field actually has two major traditions. The tradition that has been discussed most prominently and has been most influential is that of the French psychologist Alfred Binet (1857–1911).

An earlier tradition, and one that still shows some influence upon the field, is that of the English scientist Sir Francis Galton. Building on ideas put forth by his uncle Charles Darwin in *On the Origin of Species* (1859), Galton believed that human capabilities could be understood through scientific investigation. From 1884 to 1890 Galton maintained a laboratory in London where visitors could have themselves measured on a variety of psychophysical tasks, such as weight discrimination and sensitivity to musical pitch. Galton believed that psychophysical abilities were the basis of intelligence and, hence, that these tests were measures of intelligence. The earliest formal intelligence tests, therefore, required a person to perform such simple tasks as deciding which of two weights was heavier or showing how forcefully one could squeeze one's hand.

The Galtonian tradition was taken to the USA by the American psychologist James McKeen Cattell. Later, one of Cattell's students, the American anthropologist Clark Wissler, collected data showing that scores on Galtonian types of tasks were not good predictors of grades in college or even of scores on other tasks. Catell nonetheless continued to develop his Galtonian approach in psychometric research and, with Edward Thorndike, helped to establish a centre for mental testing and measurement.

The IQ Test

The more influential tradition of mental testing was developed by Binet and his collaborator, Theodore Simon, in France. In 1904 the minister of public instruction in Paris named a commission to study or create tests that would ensure that intellectually disabled children received an adequate education. The minister was also concerned that children of normal intelligence were being placed in classes for intellectually disabled children because of behaviour problems. Even before Wissler's research, Binet, who was charged with developing the new test, had flatly rejected the Galtonian tradition, believing that Galton's tests measured trivial abilities. He proposed instead that tests of intelligence should measure skills such as judgement, comprehension, and reasoning – the same kinds of skills measured by most intelligence tests today. Binet's early test was taken to Stanford University by Lewis Terman, whose version came to be called the Stanford-Binet test. This test has been revised frequently and continues to be used in countries all over the world.

The Stanford-Binet test, and others like it, have yielded at the very least an overall score referred to as an intelligence quotient, or IQ. Some tests, such as the Wechsler Adult Intelligence Scale (Revised) and the Wechsler Intelligence Scale for Children (Revised), yield an overall IQ as well as separate IQs for verbal and performance subtests. An example of a verbal subtest would be vocabulary, whereas an example of a performance subtest would be picture arrangement, the latter requiring an examinee to arrange a set of pictures into a sequence so that they tell a comprehensible story.

Later developments in intelligence testing expanded the range of abilities tested. For example, in 1997 the psychologists J.P. Das and Jack A. Naglieri published the Cognitive Assessment System, a test based on a theory of intelligence first proposed by the Russian psychologist Alexander Luria. The test measured planning abilities, attentional abilities, and simultaneous and successive processing abilities. Simultaneous processing abilities are used to solve tasks such as figural matrix problems, in which the test taker must fill in a matrix with a missing geometric form. Successive processing abilities are used in tests such as digit span, in which one must repeat back a string of memorized digits.

IQ was originally computed as the ratio of mental age to chronological (physical) age, multiplied by 100. Thus, if a child of age 10 had a mental age of 12 (that is, performed on the test at the level of an average 12-year-old), the child was assigned an IQ of $^{12}/_{10}$ 100, or 120. If the 10-year-old had a mental age of eight, the child's IQ would be $^{8}/_{10}$ 100, or 80. A score of 100, where the mental age equals the chronological age, is average.

As discussed above, the concept of mental age has fallen into disrepute. Many tests still yield an IQ, but they are most often computed on the basis of statistical distributions. The scores are assigned on the basis of what percentage of people of a given group would be expected to have a certain IQ.

The Work of Jean Piaget

The landmark work in intellectual development in the twentieth century derived not from psychometrics but from the tradition established by the Swiss psychologist Jean Piaget. His theory was concerned with the mechanisms by which intellectual development takes place and the periods through which children develop. Piaget believed that the child explores the world and observes regularities and makes generalizations – much as a scientist does. Intellectual development, he argued, derives from two cognitive processes that work in somewhat reciprocal fashion. The first, which he called assimilation, incorporates new information into an already existing cognitive structure. The second, which he called accommodation, forms a new cognitive structure into which new information can be incorporated.

The process of assimilation is illustrated in simple problem-solving tasks. Suppose that a child knows how to solve problems that require calculating a percentage of a given number. The child then learns how to solve problems that ask what percentage of a number another number is. The child already has a cognitive structure, or what Piaget called a "schema", for percentage problems and can incorporate the new knowledge into the existing structure.

Suppose that the child is then asked to learn how to solve time–rate–distance problems, having never before dealt with this type of problem. This would involve accommodation – the formation of a new cognitive structure. Cognitive development, according to Piaget, represents a dynamic equilibrium between the two processes of assimilation and accommodation.

As a second part of his theory, Piaget postulated four major periods in individual intellectual development. The first, the sensorimotor period, extends from birth to around the age of two. During this period, a child learns how to modify reflexes to make them more adaptive, to coordinate actions, to retrieve hidden objects, and, eventually, to begin representing information mentally. The second period, known as pre-operational, runs approximately from age two to age seven. In this period a child develops language and mental imagery and learns to focus on single perceptual dimensions, such as colour and size. The third, the concrete-operational period, ranges from about age seven to age 12. During this time children develop so-called conservation skills, which enable them to recognize that things that may appear to be different are actually the same – that is, that their fundamental properties are "conserved". For example, suppose that water is poured from a wide short beaker into a tall narrow one. A pre-operational child, asked which beaker has more water, will say that the second beaker does (the tall thin one); a concrete-operational child, however, will recognize that the amount of water in the beakers must be the same. Finally, children emerge into the fourth, formal-operational period, which begins at about age 12 and continues throughout life. The formal-operational child develops thinking skills in all logical combinations and learns to think with abstract concepts. For example, a child in the concrete-operational period will have great difficulty determining all the possible orderings of four digits, such as 3-7-5-8. The child who has reached the formal-operational stage, however, will adopt a strategy of systematically vary-ing alternations of digits, starting perhaps with the last digit and working toward the first. This systematic way of thinking is not normally possible for those in the concrete-operational period.

Piaget's theory had a major impact on the views of intellectual development, but it is not as widely accepted today as it was in the mid-twentieth century. One shortcoming is that the theory deals primarily with scientific and logical modes of thought, thereby neglecting aesthetic, intuitive, and other modes. In addition, Piaget erred in that children were for the most part capable of performing mental operations earlier than the ages at which he estimated they could perform them.

Despite its diminished influence, Piaget's theory continues to serve as a basis for other views. One theory has expanded on Piaget's work by suggesting a possible fifth, adult, period of development, such as "problem finding". Problem finding comes before problem solving; it is the process of identifying problems that are worth solving in the first place. A second course has identified periods of development that are quite different from those suggested by Piaget. A third course has been to accept the periods of development Piaget proposed but to hold that they have different cognitive bases. Some of the theories in the third group emphasize the importance of memory capacity. For example, it has been shown that children's difficulties in solving transitive inference problems such as: "if A is greater than B, B is greater than C, and D is less than C, which is the greatest?" result primarily from memory limitations rather than reasoning limitations (as Piaget had argued). A fourth course has been to focus on the role of knowledge in development. Some investigators argue that much of what has been attributed to reasoning and problem-solving ability in intellectual development is actually better attributed to the extent of the child's knowledge.

Artificial Intelligence

Artificial intelligence (AI) is the ability of a digital computer or computer-controlled robot to perform tasks commonly associated with intelligent beings. The term is frequently applied to the project of developing systems endowed with the intellectual processes characteristic of humans, such as the ability to reason, discover meaning, generalize, or learn from past experience. Since the development of the digital computer in the 1940s, it has been demonstrated that computers can be programmed to carry out very complex tasks – as, for example, discovering proofs for mathematical theorems or playing chess – with great proficiency.

Still, despite continuing advances in computer processing speed and memory capacity, there are as yet no programs that can match human flexibility over wider domains or in tasks requiring much everyday knowledge. On the other hand, some programs have attained the performance levels of human experts and professionals in performing certain specific tasks, so that AI in this limited sense is found in applications as diverse as medical diagnosis, computer search engines, and voice or handwriting recognition. Here are just a few of the tests that experts have devised.

Learning

There are a number of different forms of learning as applied to AI. The simplest is learning by trial and error. For example, a simple computer program for solving mate-in-one chess problems might try moves at random until mate is found. The program might then store the solution with the position so that the next time the computer encountered the same position it would recall the solution. This simple memorizing of individual items and procedures – known as rote learning – is relatively easy to implement on a computer. More challenging is the problem of implementing what is called generalization. Generalization involves applying past experience to analogous new situations. For example, a program that learns the past tense of regular English verbs by rote will not be able to produce the past tense of a word such as *jump* unless it previously had been presented with *jumped*, whereas a program that is able to generalize can learn the "add *ed*" rule and so form the past tense of *jump* based on experience with similar verbs.

Reasoning

To reason is to draw inferences appropriate to the situation. Inferences are classified as either deductive or inductive. An example of the former is, "Fred must be in either the museum or the cafe. He is not in the cafe; therefore he is in the museum," and of the latter, "Previous accidents of this sort were caused by instrument failure; therefore this accident was caused by instrument failure." The most significant difference between these forms of reasoning is that in the deductive case the truth of the premises guarantees the truth of the conclusion, whereas in the inductive case the truth of the premise lends support to the conclusion without giving absolute assurance. Inductive reasoning is common in science, where data are collected and tentative models are developed to describe and predict future behaviour – until the appearance of anomalous data forces the model to be revised. Deductive reasoning is common in mathematics and logic, where elaborate structures of irrefutable theorems are built up from a small set of basic axioms and rules.

There has been considerable success in programming computers to draw inferences, especially deductive inferences.

However, true reasoning involves more than just drawing inferences; it involves drawing inferences *relevant* to the solution of the particular task or situation. This is one of the hardest problems confronting AI.

Problem Solving

Problem solving, particularly in AI, may be characterized as a systematic search through a range of possible actions in order to reach some predefined goal or solution. Problem-solving methods divide into special purpose and general purpose. A special-purpose method is tailor-made for a particular problem and often exploits very specific features of the situation in which the problem is embedded. In contrast, a general-purpose method is applicable to a wide variety of problems. One general-purpose technique used in AI is means–end analysis – a step-by-step, or incremental, reduction of the difference between the current state and the final goal. The program selects actions from a list of means – in the case of a simple robot this might consist of PICKUP, PUTDOWN, MOVEFORWARD, MOVEBACK, MOVELEFT, and MOVERIGHT – until the goal is reached.

Many diverse problems have been solved by AI programs. Some examples are finding the winning move (or sequence of moves) in a board game, devising mathematical proofs, and manipulating "virtual objects" in a computer-generated world.

Perception

In perception the environment is scanned by means of various sensory organs, real or artificial, and the scene is decomposed into separate objects in various spatial relationships. Analysis is complicated by the fact that an object may appear different depending on the angle from which it is viewed, the direction and intensity of illumination in the scene, and how much the object contrasts with the surrounding field.

At present, artificial perception is sufficiently well advanced to enable optical sensors to identify individuals, autonomous vehicles to drive at moderate speeds on the open road, and robots to roam through buildings collecting empty soda cans. One of the earliest systems to integrate perception and action was FREDDY, a stationary robot with a moving television eye and a pincer hand, constructed at the University of Edinburgh, Scotland, during the period 1966–73 under the direction of Donald Michie. FREDDY was able to recognize a variety of objects and could be instructed to assemble simple artifacts, such as a toy car, from a random heap of components.

Language

A language is a system of signs having meaning by convention. In this sense, language need not be confined to the spoken word. Traffic signs, for example, form a mini-language, it being a matter of convention that hazard means "hazard ahead" in some countries. It is distinctive of languages that linguistic units possess meaning by convention, and linguistic meaning is very different from what is called natural meaning, exemplified in statements such as, "Those clouds mean rain" and "The fall in pressure means the valve is malfunctioning."

An important characteristic of fully-fledged human languages – in contrast to birdcalls and traffic signs – is their productivity. A productive language can formulate an unlimited variety of sentences.

It is relatively easy to write computer programs that seem able, in severely restricted contexts, to respond fluently in a human language to questions and statements. Although none of these programs actually understands language, they may, in principle, reach the point where their command of a language is indistinguishable from that of a normal human. What, then, is involved in genuine understanding, if even a computer that uses language like a native human speaker is not acknowledged to understand? There is no universally agreed upon answer to this difficult question. According to one theory, whether or not one understands depends not only on one's behaviour but also on one's history: in order to be said to understand, one must have learned the language and have been trained to take one's place in the linguistic community by means of interaction with other language users.

Symbolic vs. Connectionist Approaches

AI research follows two distinct, and to some extent competing, methods, the symbolic (or "top-down") approach, and the connectionist (or "bottom-up") approach. The top-down approach seeks to replicate intelligence by analysing cognition independent of the biological structure of the brain, in terms of the processing of symbols – whence the *symbolic* label. The bottom-up approach, on the other hand, involves creating artificial neural networks in imitation of the brain's structure – whence the *connectionist* label.

To illustrate the difference between these approaches, consider the task of building a system, equipped with an optical scanner, which recognizes the letters of the alphabet. A bottom-up approach typically involves training an artificial neural network by presenting letters to it one by one, gradually improving performance by "tuning" the network. (Tuning adjusts the responsiveness of different neural pathways to different stimuli.) In contrast, a top-down approach typically involves writing a computer program that compares each letter with geometric descriptions. Simply put, neural activities are the basis of the bottom-up approach, while symbolic descriptions are the basis of the top-down approach.

In *The Fundamentals of Learning* (1932), Edward Thorndike, a psychologist at Columbia University, New York City, first suggested that human learning consists of some unknown property of connections between neurons in the brain. In *The Organization of Behavior* (1949), Donald Hebb, a psychologist at McGill University, Montreal, Canada, suggested that learning specifically involves strengthening certain patterns of neural activity by increasing the probability (weight) of induced neuron firing between the associated connections. The notion of weighted connections is described below (page 330).

In 1957 two vigorous advocates of symbolic AI – Allen Newell, a researcher at the RAND Corporation, Santa Monica, California, and Herbert Simon, a psychologist and computer scientist at Carnegie Mellon University, Pittsburgh, Pennsylvania – summed up the top-down approach in what they called the physical symbol system hypothesis. This hypothesis states that processing structures of symbols is sufficient, in principle, to produce AI in a digital computer and that, moreover, human intelligence is the result of the same type of symbolic manipulations.

During the 1950s and 1960s, the top-down and bottom-up approaches were pursued simultaneously, and both achieved noteworthy, if limited, results. During the 1970s, however, bottom-up AI was neglected, and it was not until the 1980s that this approach again became prominent. Nowadays both approaches are followed, and both are acknowledged as facing difficulties. Symbolic techniques work in simplified realms but typically break down when confronted with the real world; meanwhile, bottom-up researchers have been unable to replicate the nervous systems of even the simplest living things. *Caenorhabditis elegans*, a much-studied worm, has approximately 300 neurons whose pattern of interconnections is perfectly known. Yet connectionist models have failed to mimic even this worm. Evidently, the neurons of connectionist theory are gross oversimplifications of the real thing.

Strong AI, Applied AI, and Cognitive Simulation

Employing the methods outlined above, AI research attempts to reach one of three goals: strong AI, applied AI, or cognitive simulation. Strong AI aims to build machines that think. (The term "strong AI" was introduced for this category of research in 1980 by the philosopher John Searle, of the University of California at Berkeley.) The ultimate ambition of strong AI is to produce a machine whose overall intellectual ability is indistinguishable from that of a human being. As is described on page 168, this goal generated great interest in the 1950s and '60s, but such optimism has given way to an appreciation of the extreme difficulties involved. To date, progress has been meagre. Some critics doubt whether research will produce even a system with the overall intellectual ability of an ant in the forseeable future. Indeed, some researchers working in AI's other two branches view strong AI as not worth pursuing.

Applied AI, also known as advanced information processing, aims to produce commercially viable "smart" systems, e.g. "expert" medical diagnosis systems and stock-trading systems. In cognitive simulation, computers are used to test theories about how the human mind works – for example, theories about how people recognize faces or recall memories. Cognitive simulation is already a powerful tool in both neuroscience and cognitive psychology.

The Turing Test

The earliest substantial work in the field of AI was done in the mid-twentieth century by the British logician and computer pioneer Alan Mathison Turing. In 1935 Turing described an abstract computing machine consisting of a limitless memory and a scanner that moves back and forth through the memory, symbol by symbol, reading what it finds and writing further symbols. The actions of the scanner are dictated by a program of instructions that also is stored in the memory in the form of symbols. This is Turing's stored-program concept, and implicit in it is the possibility of the machine operating on, and so modifying or improving, its own program. Turing's conception is now known simply as the universal Turing machine. All modern computers are in essence universal Turing machines.

Turing gave quite possibly the earliest public lecture (London, 1947) to mention computer intelligence, saying, "What we want is a machine that can learn from experience," and that the "possibility of letting the machine alter its own instructions provides the mechanism for this". In 1948 he introduced many of the central concepts of AI in a report entitled "Intelligent Machinery". However, Turing did not publish this paper, and many of his ideas were later reinvented by others. For instance, one of Turing's original ideas was to train a network of artificial neurons to perform specific tasks, an approach described on page 171.

In 1991 the American philanthropist Hugh Loebner started the annual Loebner Prize competition, promising a US$100,000 payout to the first computer to pass the Turing test and awarding $2,000 each year to the best effort. However, no AI program has come close to passing an undiluted Turing test.

The Brain as a Computer

Connectionism, or neuron-like computing, developed out of attempts to understand how the human brain works at the neural level and, in particular, how people learn and remember. In 1943 the neurophysiologist Warren McCulloch of the University of Illinois and the mathematician Walter Pitts of the University of Chicago published an influential treatise on neural nets and automatons, according to which each neuron in the brain is a simple digital processor and the brain as a whole is a form of computing machine. As McCulloch put it subsequently, "What we thought we were doing (and I think we succeeded fairly well) was treating the brain as a Turing machine."

Recent work on neuron-like computing includes the following:

Visual perception. Networks can recognize faces and other objects from visual data. A neural network designed by John Hummel and Irving Biederman at the University of Minnesota can identify about 10 objects from simple line drawings. The network is able to recognize the objects – which include a mug and a frying pan – even when they are drawn from different angles. Networks investigated by Tomaso Poggio of MIT are able to recognize bent-wire shapes drawn from different angles, faces photographed from different angles and showing different expressions, and objects from cartoon drawings with grey-scale shading indicating depth and orientation.

Language processing. Neural networks are able to convert handwritten and typewritten material to electronic text. The US Internal Revenue Service has commissioned a neuron-like system that will automatically read tax returns and correspondence. Neural networks also convert speech to printed text and printed text to speech.

Financial analysis. Neural networks are being used increasingly for loan risk assessment, real estate valuation, bankruptcy prediction, share price prediction, and other business applications.

Medicine. Medical applications include detecting lung nodules and heart arrhythmias and predicting adverse drug reactions.

Telecommunications. Telecommunications applications of neural networks include control of telephone switching networks and echo cancellation in modems and on satellite links.

8

THE SCIENCE OF SLEEP

Sleep remains one of the great mysteries of modern neuroscience. Humans spend nearly one-third of their lives asleep, but the function of sleep still is not known. Fortunately, over the past few years researchers have made great headway in understanding some of the brain circuitry that controls wake–sleep states.

Sleep is a normal, easily reversible, recurrent, and spontaneous state of decreased and less-efficient responsiveness to external stimulation. The state contrasts with that of wakefulness, in which there is an enhanced potential for sensitivity and an efficient responsiveness to external stimuli. The sleep–wakefulness alternation is the most striking manifestation in higher vertebrates of the more general phenomenon of periodicity in the activity or responsivity of living tissue.

There is no single, perfectly reliable criterion of sleep. Sleep is defined by the convergence of observations satisfying several different motor, sensory, and physiological criteria. Occasionally, one or more of these criteria may be absent during sleep or present during wakefulness, but even in such cases there usually is little difficulty in achieving agreement among observers in the discrimination between the two behavioural states.

Sleep usually requires the presence of flaccid or relaxed skeletal muscles and the absence of the overt, goal-directed behaviour of which the waking organism is capable. Part of the recurring fascination with sleep talking and sleepwalking stems from their apparent violation of this latter criterion. Were these phenomena continuous rather than intermittent during a behavioural state, it is indeed questionable whether the designation "sleep" would continue to be appropriate. The characteristic posture associated with sleep in humans and in many but not all other animals is that of horizontal repose. The relaxation of the skeletal muscles in this posture and its implication of a more passive role toward the environment are symptomatic of sleep.

Indicative of the decreased sensitivity of the human sleeper to their external environment are the typical closed eyelids (or the functional blindness associated with sleep while the eyes are open) and the presleep activities that include seeking surroundings characterized by reduced or monotonous levels of sensory stimulation. Three additional criteria – reversibility, recurrence, and spontaneity – distinguish the insensitivity of sleep from that of other states. Compared with that of hibernation or coma, the insensitivity of sleep is more easily reversible. Although the occurrence of sleep is not perfectly regular under all conditions, it is at least partially predictable from a knowledge of the duration of prior sleep periods and of the intervals between periods of sleep, and, although the onset of sleep may be facilitated by a variety of environmental or chemical means, sleep states are not thought of as being absolutely dependent upon such manipulations.

In experimental studies, both with subhuman vertebrates and with humans, sleep also has been defined in terms of physiological variables generally associated with recurring periods of inactivity identified behaviourally as sleep. For example, the typical presence of certain EEG patterns (brain patterns of electrical activity as recorded in tracings) with behavioural sleep has led to the designation of such patterns as "signs" of sleep. Conversely, in the absence of such signs (as, for example, in a hypnotic trance), it is felt that true sleep is absent. Such signs as are now employed, however, are not invariably discriminating of the behavioural states of sleep and wakefulness. Advances in the technology of animal experimentation have made it possible to extend the physiological approach from externally measurable manifestations of sleep such as the EEG to the underlying neural (nerve) mechanisms presumably responsible for such manifestations. As a result, it may finally become possible to identify structures or functions that are invariably related to behavioural sleep and to trace the evolution of sleep through comparative anatomical and physiological studies of structures found to be critical in the maintenance of sleep behaviour in the higher vertebrates.

In addition to the behavioural and physiological criteria already mentioned, subjective experience (in the case of the self) and verbal reports of such experience (in the case of others) are used at the human level to define sleep. Upon being alerted, one may feel or say, "I was asleep just then", and such judgements ordinarily are accepted as evidence for identifying a pre-arousal state as sleep, but such subjective evidence can be at variance with behaviouristic classifications of sleep.

More generally, problems in defining sleep arise when evidence for one or more of the several criteria of sleep is lacking or when the evidence generated by available criteria is inconsistent. Do subhuman species sleep? Other mammalian species whose EEG and other physiological correlates are akin to those observed in human sleep demonstrate recurring, spontaneous, and reversible periods of inactivity and decreased critical reactivity. There is general acceptance of the designation of such states as sleep. As one descends the evolutionary scale below the birds and reptiles, however, and such criteria are successively less well satisfied, the unequivocal identification of sleep becomes more difficult. Bullfrogs (*Rana catesbeiana*), for example, seem not to fulfil sensory threshold criteria of sleep during resting states. Tree frogs (genus *Hyla*), on the other hand, show diminished sensitivity as they move from a state of behavioural activity to one of rest. Yet the EEGs of the alert rest of the bullfrog and the sleep-like rest of the tree frog are the same. There are parallel problems in defining sleep at different stages in the development of a single individual. At full-term birth in the human, for instance, a convergence of non-subjective criteria clearly seems to justify the identification of periods of sleep, but it is more difficult to justify the attribution of sleep to the human fetus.

Problems in defining sleep may arise from the effects of artificial manipulation. For example, the EEG patterns commonly used as signs of sleep can be induced in an otherwise waking organism by the administration of certain drugs. Sometimes, also, there is conflicting evidence: someone who is "awakened" from a spontaneously assumed state of immobility with all the EEG criteria of sleep may claim that they had been awake prior to this event. In such troublesome cases and more generally, it is becoming common to qualify attributions of sleep with the criteria upon which such attributions rest, e.g. "behavioural sleep", "physiological sleep", or "self-described sleep". Such terminology accurately reflects both the multiplicity of criteria available for the identification of sleep and the possibility that these criteria may not always agree with one another.

Theories of Sleep

Two kinds of sleep theories of contemporary interest may be distinguished. One begins with the peripheral physiology of sleep and relates it to underlying neural (nervous system) or biochemical mechanisms. Such theories most often rely on experiments with animals by means of drugs or surgery. Alternatively, sleep theories may start with behavioural observations of sleep and may attempt to specify the functions of such a state of lethargy and insensitivity from an evolutionary or adaptive point of view. The question here is not so much how people sleep, or even why they sleep, but what good it does.

Mechanistic Theories

Historically, mechanistic theories of sleep have focused on a succession of organs or structures in a manner reflective of the degree of access different civilizations have had to the inner workings of the human body. Thus, the relatively perceptible processes of circulation, digestion, and secretion played large roles in the theories of classical antiquity, and modern theories have been concerned with the central nervous system, particularly the brain, although various peripheral factors in the induction of sleep have not been ruled out. Proposals that blood composition, metabolic changes, or internal secretions regulate sleep are necessarily incomplete to the extent that they ignore the contributions of environment and intent to the onset of sleep. It also has been noted that, in humans born with two heads, one "twin" may seem asleep while the other is awake, despite their sharing a circulatory system.

Neural Theories

Among neural theories of sleep, there are certain issues that each must face. Is the sleep–wakefulness alternation to be considered a property of individual neurons (nerve cells), making unnecessary the postulation of specific regulative centres, or is it to be assumed that there are some aggregations of neurons that play a dominant role in sleep induction and maintenance? The Russian physiologist Ivan Petrovich Pavlov adopted the former position, proposing that sleep is the result of irradiating inhibition among cortical and subcortical neurons (nerve cells in the outer brain layer and in the brain layers beneath the cortex). Microelectrode studies, on the other hand, have revealed high rates of discharge during sleep from many neurons in the motor and visual areas of the cortex, and it thus seems that, as compared with wakefulness, sleep must consist of a different organization of cortical activity rather than some general, overall decline.

Another issue has been whether there is a waking centre, fluctuations in whose level of functioning are responsible for various degrees of wakefulness and sleep, or whether the induction of sleep requires another centre, actively antagonistic to the waking centre. Early speculation favoured the passive view of sleep. A *cerveau isolé* preparation, an animal in which a surgical incision high in the midbrain has separated the cerebral hemispheres from sensory input, demonstrated chronic somnolence. It has been reasoned that a similar cutting off of sensory input, functional rather than structural, must characterize natural states of sleep. Other supporting observations for the stimulus-deficiency theory of sleep included pre-sleep rituals such as turning out the lights, regulation of stimulus input, and the facilitation of sleep induction by muscular relaxation. With the discovery of the ascending reticular activating system (ARAS; a network of nerves in the brainstem), it was found that it is not the sensory nerves themselves that maintain cortical arousal but rather the ARAS, which projects impulses diffusely to the cortex from the brainstem. Presumably, sleep would result from interference with the active functioning of the ARAS. Injuries to the ARAS were in fact found to produce sleep. Sleep thus seemed passive, in the sense that it was the absence of something (ARAS support of sensory impulses) characteristic of wakefulness.

Theory has tended to depart from this belief and to move toward conceiving of sleep as an actively produced state. Two kinds of observation primarily have been responsible for the shift. First, earlier studies showing that sleep can be induced directly by electrical stimulation of certain areas in the hypothalamus have been confirmed and extended to other areas in the brain. Second, the discovery of rapid eye movement (REM) sleep has been even more significant in leading theorists to consider the possibility of actively produced sleep. REM sleep, by its very active nature, defies description as a passive state. As is noted below, REM sleep can be eliminated in experimental animals by the surgical destruction of a group of nerve cells in the pons, the active function of which appears to be necessary for REM sleep. Thus, it is difficult to imagine that the various manifestations of REM sleep reflect merely the deactivation of wakefulness mechanisms.

The rapid eye movement–non-rapid eye movement (REM–NREM) sleep dichotomy poses a third issue for the theories of sleep mechanisms, or at least for those who accept the idea of sleep as an active phenomenon. Does one hypnogenic (sleep-causing) system serve both kinds of sleep, or are there two antagonistic sleep systems, one for REM sleep and one for NREM sleep? Opinion is sharply divided. One group of theorists states that there must be two sleep systems. It is noted that NREM sleep is not affected – but REM sleep is abolished – by injuries to the pontine tegmentum (the posterior part of the pons) and that NREM sleep is suppressed in animals whose brainstem has been severed at the midpoint of the pons, suggesting that an NREM-sleep centre behind this section no longer is capable of suppressing the effect of the ARAS. It is further observed that the neurohumour serotonin is localized in the brainstem regions presumed to be responsible for NREM sleep; that destruction of serotonin-containing nerve cells in the brainstem may produce insomnia; that, in some species, reductions of serotonin by chemical interference with its production produces an amount of sleep loss correlated with the reduction of serotonin; that administration of a serotonin precursor (a substance from which serotonin is formed) after interference with production of serotonin produces a sleep-like state and that artificially induced increases in brain serotonin increase NREM sleep; that the neurohumour norepinephrine is localized in the brainstem regions presumed to be responsible for REM sleep; and that substances interfering with the synthesis of norepinephrine suppress REM sleep. Other theorists have proposed that REM and NREM sleep are served by a common hypnogenic system. Chemical stimulation of certain brain structures, assumed to constitute a hypnogenic system, has been found capable of inducing both stages of sleep. It also is argued that different varieties of sleep should require different mechanisms no more than do different varieties of wakefulness (e.g. alertness, relaxation).

Functional Theories

Functional theories stress the recuperative and adaptive value of sleep. Sleep arises most unequivocally in animals that maintain a constant body temperature and that can be active at a wide range of environmental temperatures. In such forms, increased metabolic requirements may find partial compensation in periodic decreases in body temperature and metabolic rate (i.e. during NREM sleep). Thus, the parallel evolution of temperature regulation and NREM sleep has suggested to some authorities that NREM sleep may best be viewed as a regulatory mechanism conserving energy expenditure in species whose metabolic requirements are otherwise high. As a solution to the problem of susceptibility to predation that comes with the torpor of sleep, it has been suggested that the periodic reactivation of the organism during sleep better prepares it for fight or flight and that the possibility of enhanced processing of significant environmental stimuli during REM sleep may even reduce the need for sudden confrontation with danger. Other functional theorists agree that NREM sleep may be a state of "bodily repair", while suggesting that REM sleep is one of "brain repair" or restitution, a period, for example, of increased cerebral protein synthesis or of "reprogramming" the brain so that information achieved in wakeful functioning is most efficiently assimilated. In their specification of functions and provision of evidence for such functions, such theories are necessarily vague and incomplete. The function of stage two NREM sleep is still unclear, for example. Such sleep is present in only rudimentary form in subprimate species yet consumes approximately half of human sleep time. Comparative, physiological, and experimental evidence is unavailable to suggest why so much human sleep is spent in this stage. In fact, poor sleepers whose laboratory sleep records show high proportions of stage two and little or no REM sleep often report feeling they have not slept at all.

How Much Sleep Does a Person Need?

While the physiological bases of the need for sleep remain conjectural, rendering definitive answers to this question impossible, much evidence has been gathered on how much sleep people do in fact obtain. Perhaps the most important conclusion to be drawn from this evidence is that there is great variability between individuals in total sleep time. For adults, anything between six and nine hours of sleep as a nightly average is not unusual, and seven hours probably best expresses the norm. Such norms, of course, inevitably vary with the criteria of sleep employed. The most precise and reliable figures on sleep time, including those cited here, come from studies in sleep laboratories, where EEG criteria are employed.

Age consistently has been associated with the varying amount, quality, and patterning of electrophysiologically defined sleep. The newborn infant may spend an average of about 16 hours of each 24-hour period in sleep, although there is wide variability between individual babies. During the first year of life, total sleep time drops sharply; by two years of age, it may range from nine to 12 hours. Decreases to approximately six hours have been observed among the elderly.

As will be elaborated below, EEG sleep studies have indicated that sleep can be considered to consist of several different stages. Developmental changes in the relative proportion of sleep time spent in these sleep stages are as striking as age-related changes in total sleep time. For example, the newborn infant may spend 50 per cent of total sleep time in a stage of EEG sleep that is accompanied by intermittent bursts of rapid eye movements (REMs) indicative of a type of sleep that in some respects bears more resemblance to wakefulness than to other forms of sleep (see page 186), while the comparable figure for adults is approximately 25 per cent and for the elderly is less than 20 per cent. There is also a decline with age of EEG stage four (deep slumber).

Sleep patterning consists of (1) the temporal spacing of sleep and wakefulness within a 24-hour period and (2) the ordering of different sleep stages within a given sleep period. In both senses there are major developmental changes in the patterning of sleep. In alternations between sleep and wakefulness, there is a developmental shift from polyphasic sleep to monophasic sleep (i.e. from intermittent to uninterrupted sleep). At birth there may be five or six periods of sleep per day alternating with a like number of waking periods. With the dropping of nocturnal feedings in infancy and of morning and afternoon naps in childhood, there is an increasing tendency to the concentration of sleep in one long nocturnal period. The trend to monophasic sleep probably reflects some blend of the effects of maturing and of pressures from a culture geared to daytime activity and nocturnal rest. Among the elderly there may be a partial return to the polyphasic sleep pattern of infancy and early childhood, namely, more frequent daytime napping and less extensive periods of nocturnal sleep because of the loss of zeitgebers, or time markers that provide cues. These include the need to arise at a set time for work or to get children off to school. Significant developmental effects also have been observed in spacing of stages within sleep. In the adult, REM sleep rarely occurs at sleep onset, while, in newborn infants, sleep-onset REM sleep is typical.

It would be difficult to overestimate the significance of the various age-related changes in sleep behaviour for a general theory of sleep. In the search for the functional significance of sleep or of particular stages of sleep, the shifts in sleep variables can be linked with variations in waking developmental needs, the total capacities of the individual, and environmental demands. It has been suggested, for instance, that the high frequency and priority in the night of REM sleep in the newborn infant may reflect a need for stimulation from within to permit orderly maturation of the central nervous system. Another interpretation of age-related changes in REM sleep stresses its possible role in processing new information, the rate of acquisition for which is assumed to be relatively high in childhood but reduced in old age. As these views illustrate, developmental changes in the electrophysiology of sleep are germane not only to sleep but also to the role of central nervous system development in behavioural adaptation.

Counting Sheep

That there are different kinds of sleep has long been recognized. In everyday discourse there is talk of "good" sleep or "poor" sleep, of "light" sleep and "deep" sleep; yet not until the second half of the twentieth century did scientists pay much attention to qualitative variations within sleep. Sleep was formerly conceptualized by scientists as a unitary state of passive recuperation. Revolutionary changes have occurred in scientific thinking about sleep, the most important of which has been increased sensitivity to its heterogeneity.

This revolution may be traced back to the discovery of sleep characterized by REM sleep, first reported by the physiologists Eugene Aserinsky and Nathaniel Kleitman in 1953. REM sleep proved to have characteristics quite at variance with the prevailing model of sleep as recuperative deactivation of the central nervous system. Various central and autonomic nervous system measurements seemed to show that the REM stage of sleep is more nearly like activated wakefulness than it is like other sleep. It now has become conventional to consider REM ("paradoxical") and non-REM (NREM or "orthodox") sleep as qualitatively different. Thus, the earlier assumption that sleep is a unitary and passive state has yielded to the viewpoint that there are two different kinds of sleep, a relatively deactivated NREM phase and an activated REM phase.

NREM sleep itself is conventionally subdivided into several different stages on the basis of EEG criteria. In the adult, stage one is observed at sleep onset or after momentary arousals during the night and is defined as a low-voltage mixed-frequency EEG tracing with a considerable representation of theta-wave (4–7 Hz, or cycles per second) activity. Stage two is a relatively low-voltage EEG tracing characterized by intermittent, short sequences of waves of 12–14 Hz ("sleep spindles") and by formations called K-complexes – biphasic wave forms that can be induced by external stimulation, as by a sound, but that also occur spontaneously during sleep. Stages three and four consist of relatively high-voltage (more than 50 ìV) EEG tracings with a predominance of delta-wave (1–2 Hz) activity; the distinction between the two stages is based on an arbitrary criterion of amount of delta-wave activity, with greater amounts classified as stage four. Unlike the basic distinction between NREM and REM, differences between NREM sleep stages generally are regarded as quantitative rather than qualitative.

The EEG patterns of NREM sleep, particularly of stages three and four (tracings of slower frequency and higher amplitude), are those associated in other circumstances with decreased vigilance. Furthermore, after the transition from wakefulness to NREM sleep, most functions of the autonomic nervous system decrease their rate of activity and their moment-to-moment variability. Thus, NREM sleep is the kind of seemingly restful state that appears capable of supporting the recuperative functions assigned to sleep. There are in fact several lines of evidence suggesting such functions for NREM stage four: (1) increases in such sleep, in both humans and laboratory animals, observed after physical exercise; (2) the concentration of such sleep in the early portion of the sleep period (i.e. immediately after wakeful states of activity) in humans; and (3) the relatively high priority that such sleep has among humans in "recovery" sleep following abnormally extended periods of wakefulness.

REM sleep is a state of diffuse bodily activation. Its EEG patterns (tracings of faster frequency and lower amplitude than in NREM stages two to four) are at least superficially similar to those of wakefulness. Most autonomic variables exhibit relatively high rates of activity and variability during REM sleep; for example, there are higher heart and respiration rates and more short-term variability in these rates than in NREM sleep, increased blood pressure, and, in males, full or partial penile erection. In addition, REM sleep is accompanied by a relatively low rate of gross body motility, but with some periodic twitching of the muscles of the face and extremities, relatively high levels of oxygen consumption by the brain, increased cerebral blood flow, and higher brain temperature. An even more impressive demonstration of the activation of REM sleep is to be found in the firing rates of individual cerebral neurons, or nerve cells, in experimental animals: during REM sleep such rates exceed those of NREM sleep and often equal or surpass those of wakefulness. Another distinguishing feature of REM sleep of course is the intermittent appearance of bursts of rapid eye movements, whence the term is derived.

For both humans and animals, REM sleep is now defined by the concurrence of three events: low-voltage, mixed-frequency EEG; intermittent REMs; and suppressed tonus of the muscles of the facial region (i.e. suppression of the continuous slight tension otherwise normally present). This decrease in muscle tonus and a similarly observed suppression of spinal reflexes are indicative of heightened motor inhibition during REM sleep.

Animal studies have identified the locus ceruleus, in the pons, as the probable source of this inhibition. (The pons is in the brainstem, directly above the medulla oblongata; the locus ceruleus borders on the brain cavity known as the fourth ventricle.) When this structure is surgically destroyed in experimental animals, they periodically engage in active, apparently goal-directed behaviour during REM sleep, although they still show the unresponsivity to external stimulation characteristic of the stage. It has been suggested that such behaviour may be the acting out of the hallucinations of a dream.

An important theoretical distinction is that between REM sleep phenomena that are continuous and those that are intermittent. Tonic (continuous) characteristics of REM sleep include the low-voltage EEG and the suppressed muscle tonus; intermittent events in REM sleep include the REMs themselves and, as observed in the cat, spike-like electrical activity in those parts of the brain concerned with vision and in other parts of the cerebral cortex. The various intermittent events of REM sleep tend to occur together, and it seems to be these moments of intermittent activation that are responsible for much of the difference between REM sleep and NREM sleep. The spiking mentioned is observed occasionally in NREM sleep, an occurrence that has been interpreted by some theorists as suggesting that REM sleep is not qualitatively unique in its capacity to support intermittent activation and that the differences between NREM and REM sleep may be less striking than the differences in eye movement and EEG have indicated.

Sleep Deprivation

One time-honoured approach to determining the function of an organ or process is to deprive an organism of the organ or process. In the case of sleep, the deprivation approach to function has been applied – both experimentally and naturalistically – to sleep as a unitary state (general sleep deprivation) and, experimentally only, to particular kinds of sleep (selective sleep deprivation). General sleep deprivation may be either total (e.g. someone has had no sleep at all for a period of days) or partial (e.g. over a period a person obtains only three or four hours of sleep per night). The method of general deprivation studies is enforced wakefulness. Selective deprivation has been reported for two stages of sleep: stage four of NREM sleep and REM sleep. Both typically occur after the appearance of other sleep stages, REM sleep after all four NREM stages and stage four after the lighter NREM stages. The general idea of selective deprivation studies is to allow the sleeper to have natural sleep until the point at which they enter the stage to be deprived and then to prevent the stage, either by experimental awakening or by other manipulations such as application of a mildly noxious stimulus or prior administration of a drug known to suppress it. The hope is that total sleep time will not be altered but that increased occurrence of some other stage will substitute for the loss of the one selectively eliminated.

Sleep Disorders

It is important at the outset to emphasize that, as dramatic and reliable as the various stages of sleep are, their functions or relations to waking performance, mood, and health are still largely unknown. Thus, association of a sleep abnormality with a certain stage of sleep (either in the sense that an abnormal event occurs during a certain stage or in the sense that an abnormal condition is associated with an increase or decrease in the proportion of total sleep time spent in that stage) is difficult to interpret when the function or necessity of that stage is uncertain. The pathology of sleep includes (1) primary disturbances of sleep–wakefulness mechanisms, such as seem to characterize encephalitis lethargica (sleeping sickness), narcolepsy (irresistible brief episodes of sleep), and hypersomnia (sleep attacks of lesser urgency but greater duration than those of narcolepsy), (2) minor episodes occurring during sleep, such as bed-wetting and nightmares, (3) medical disorders such as sleep apnea whose symptoms occur during sleep, (4) sleep symptoms of the major psychiatric disorders, and (5) disorders of sleep schedule.

Epidemic lethargic encephalitis is produced by viral infections of sleep–wakefulness mechanisms in the hypothalamus, a structure at the upper end of the brainstem. The disease often passes through several stages: fever and delirium, hyposomnia (loss of sleep), and hypersomnia (excessive sleep, sometimes bordering on coma). Inversions of 24-hour sleep–wakefulness patterns are also commonly observed, as are disturbances in eye movements.

Narcolepsy, like encephalitis, is thought to involve specific abnormal functioning of subcortical sleep-regulatory centres. Some people who experience attacks of narcolepsy also have one or more of the following auxiliary symptoms: cataplexy, a sudden loss of muscle tone often precipitated by an emotional response such as laughter or startle and sometimes so dramatic as to cause the person to fall down; hypnagogic (sleep onset) and hypnopompic (awakening) visual hallucinations of a dream-like sort; and hypnagogic or hypnopompic sleep paralysis, in which the person is unable to move voluntary muscles (except respiratory muscles) for a period ranging from several seconds to several minutes. When narcolepsy includes one or more of these accessory symptoms, some of the sleep attacks consist of periods of REM at the onset of sleep. This precocious triggering of REM sleep (which occurs in adults generally only after 70–90 minutes of NREM sleep) may indicate that the accessory symptoms are dissociated aspects of REM sleep, i.e. the cataplexy and the paralysis represent the active motor inhibition of REM sleep, and the hallucinations represent the dream experience of REM sleep. Thus, narcolepsy involves REM sleep, and it is thought that it probably involves a failure of wakefulness mechanisms to inhibit the REM-sleep mechanisms.

Hypersomnia may involve either excessive daytime sleep and drowsiness or a nocturnal sleep period of greater than normal duration, but it does not include sleep-onset REM periods. One reported concomitant of hypersomnia, the failure of the heart rate to decrease during sleep, suggests that hypersomniac sleep may not be as restful per unit of time as is normal sleep. In its primary form, hypersomnia is probably hereditary in origin (as is narcolepsy) and is thought to involve some disruption of the functioning of hypothalamic sleep centres. Narcolepsy and hypersomnia are not characterized by grossly abnormal EEG sleep patterns. The abnormality seems to involve a failure in "turn on" and "turn off" mechanisms regulating sleep, rather than in the sleep process itself. Narcoleptic and hypersomniac symptoms can be managed by administration of drugs. Several forms of hypersomnia are periodic rather than chronic. One rare disorder of periodically excessive sleep, the Kleine–Levin syndrome, is characterized by periods of two to four weeks of excessive sleep, along with a ravenous appetite and psychotic-like behaviour during the few waking hours. The Pickwickian syndrome (in reference to Joe, the fat boy, in Charles Dickens's *The Pickwick Papers*), another form of periodically excessive sleep, is associated with obesity and respiratory insufficiency.

Hyposomnia (this word, meaning "too little sleep", is chosen in preference to "insomnia", or "lack of sleep", because some sleep invariably is present) is less clearly under-stood than the conditions already mentioned. It has been demonstrated that, by physiological criteria, self-described poor sleepers generally sleep much better than they imagine. Their sleep, however, does show signs of disturbance: frequent body movement, enhanced levels of autonomic functioning, reduced levels of REM sleep, and in some the intrusion of waking rhythms (alpha waves) throughout the various sleep stages. Although hyposomnia in a particular situation is common and without pathological import, chronic hyposom-nia may be related to psychological disturbance. Hyposomnia is conventionally treated by administration of drugs but often with substances that are potentially addictive and otherwise dangerous when used over long periods. Newer treatments involve behavioural programmes such as the temporary re-striction of sleep time and its gradual reinstatement.

Among the minor episodes sometimes considered abnormal in sleep are somniloquy (sleep talking) and somnambulism (sleepwalking), enuresis (bed-wetting), bruxism (teeth grind-ing), snoring, and nightmares. Sleep talking seems more often to consist of inarticulate mumblings than of extended, mean-ingful utterances. It occurs at least occasionally for many people and at this level cannot be considered pathological. Sleepwalking is not uncommon in children, but its continu-ation into adulthood is suggestive of persistent immaturity of the central nervous system. Enuresis may be a secondary symptom of a variety of organic conditions or, more fre-quently, a primary disorder in its own right. In the latter case, it seems to involve some immaturity in neural control of bladder muscles. While mainly a disorder of early childhood, enuresis persists into adulthood for a few people. Treatment generally has been directed either toward sensitizing the sleeper to bladder distention, so that they will awaken and urinate according to appropriate social norms, or toward increasing bladder capacity. Primary enuresis does not seem to be an abnormality of sleep, sleep cycles of bed-wetting children and of non-bed-wetting children being roughly the same. Teeth grinding is not consistently associated with any particular stage of sleep, nor does it appreciably affect overall sleep patterning; it too seems to be an abnormality in, rather than of, sleep.

A variety of frightening experiences associated with sleep have, at one time or another, been called nightmares. Because not all such phenomena have proved to be identical in their associations with sleep stages or with other variables, several distinctions need to be made between them. Incubus, the classic nightmare of adult years, consists of arousal from stage 4 NREM sleep with a sense of heaviness over the chest and with diffuse anxiety but with little or no dream recall. Night terrors (*pavor nocturnus*) are a disorder of early childhood. Delta-wave NREM sleep is suddenly interrupted; the child may scream and sit up in apparent terror and be incoherent and inconsolable. After a period of minutes, they return to sleep, often without ever having been fully alert or awake. Dream recall generally is absent, and the entire episode may be forgotten in the morning. Anxiety dreams most often seem associated with spontaneous arousals from REM sleep. There is remembrance of a dream whose content is in keeping with the disturbed awakening. While their persistent recurrence probably indicates waking psychological disturbance or stress caused by a difficult situation, anxiety dreams occur occasionally in many otherwise healthy people.

A variety of medical symptoms may be accentuated by the conditions of sleep. Attacks of angina (spasmodic, choking pain), for example, apparently can be augmented by the activation of the autonomic nervous system in REM sleep;

the same is true of gastric acid secretions in people with duodenal ulcers. NREM sleep, on the other hand, can increase the likelihood of certain kinds of epileptic discharge.

Rhythmic snoring, which can occur throughout sleep, indicates the partial muscular relaxation of sleep, and its occasional occurrence is not abnormal. When snoring is of the loud, laboured, snorting variety, however, and is accompanied by pauses in respiration of more than 10 seconds in duration, broken by gasping sounds, the respiratory disorder called sleep apnea may be present. This disorder can occur at any age but is most common in the elderly. It results in hypoxia and sleep fragmentation, both of which contribute to excessive daytime sleepiness and cognitive deficits. Treatment approaches include behaviour change (reduction of alcohol consumption and body weight), sleep-position training, mechanical appliances to keep the airway unobstructed, and surgery.

The resemblance of dream consciousness to waking psychotic experience has often been noted, and the psychotic has been considered a "waking dreamer". Thus, it has been theorized that waking psychotic symptoms may be generated by a spontaneous or REM sleep deprivation-induced shift of REM phenomena from sleep to the waking state. Symptomatically, schizophrenics have shown neither the exacerbation of psychotic symptoms under experimental REM-sleep deprivation nor the consistent or large deviations from normal EEG sleep patterning that would seem to be required by the hypothesis that sleep mechanisms play some critical role in bringing on psychotic episodes. Depressed people do sleep less and have an earlier first REM period than non-depressed people. The first REM period, occurring 40–60 minutes after sleep onset, is often longer than normal, with more eye movement activity. This suggests a disruption in the drive-regulation function, affecting such things as sexuality, appetite, or aggressiveness, all of which are reduced in such people. REM deprivation by pharmacological agents (tricyclic antidepressants) or by REM-awakening techniques appears to reverse this sleep abnormality and to relieve the waking symptoms.

There are two prominent types of sleep-schedule disorders: phase-advanced sleep and phase-delayed sleep. In the former the sleep onset and offset occur earlier than the social norms, and in the latter sleep onset is delayed and waking is also later in the day than is desirable. These alterations in the sleep–wake cycle may occur in shift workers or following international travel across time zones. They can be treated by gradual readjustment of the timing of sleep.

9

THE MIND AND CONSCIOUSNESS

The study of the philosophy of mind is the investigation of the nature of the mind and of mental acts, including such problems as the relation of the mind to the body and the knowledge of other minds.

It should be clear that the range of topics in the philosophy of mind goes far beyond what is intended in everyday discourse by "mind". When, for example, the layman speaks of someone as having "a good mind" or as pursuing "the pleasures of the mind", they are thinking of those particular activities that have to do with abstract reasoning, intellectual pursuits, and the exercise of intelligence. The "mind", as the term is used in the philosophy of mind, encompasses a variety of elements including sensation and sense perception, feeling and emotion, dreams, traits of character and personality, the unconscious, and the volitional aspects of human life, as well as the more narrowly intellectual phenomena, such as thought, memory, and belief.

Although the philosophy of mind is a distinct field of investigation, it has many important relations with other fields. First, its methods, being those of philosophy in general, are to be tested by the fruits that they have yielded in other areas: if a method has been successful in other areas, it is reasonable to try it here; if unsuccessful in other areas, it is suspect here. Second, the conclusions achieved in such fields as epistemology, metaphysics, logic, ethics, and the philosophy of religion are quite relevant to the philosophy of mind; and its conclusions, in turn, have important implications for those fields. Moreover, this reciprocity applies as well to its relations to such empirical disciplines as neurology, psychology, sociology, and history. Thus, the philosopher of mind must keep informed of developments in all related fields of investigation.

The basic metaphysical issues in the philosophy of mind concern what kind of existence the mind has and what its relation is to the rest of what exists. Historically there have been three main positions. Materialists held that only physical matter (and physical energy) exists. Dualists, on the other hand, maintained that both immaterial minds and material bodies exist, and idealists that only minds exist but not bodies. Finally, so-called neutral monists argued that the fundamental existents in nature are neither mental nor physical but some neutral stuff out of which both the mental and the material are formed.

Theories of the Mind

Since the late twentieth century, the major debate in the philosophy of mind has concerned the question of which materialist theory of the human mind, if any, is the correct one. The main theories are the identity theory (also called reductive materialism), functionalism, and eliminative materialism.

Identity Theory

An early form of identity theory held that each type of mental state, such as pain, is identical with a certain type of physical state of the human brain or central nervous system. This encountered two main objections. First, it falsely implies that only human beings can have mental states. Second, it is inconsistent with the plausible intuition that it is possible for two human beings to be in the same mental state (such as the state of believing that the King of France is bald) and yet not be in the same neurophysiological state.

As a result of these and other objections, type-type identity theory was discarded in favour of what was called "token-token" identity theory. According to this view, particular instances or occurrences of mental states, such as the pain felt by a particular person at a particular time, are identical with particular physical states of the brain or central nervous system. Even this version of the theory, however, seemed to be inconsistent with the plausible intuition that felt sensation is not identical with neural activity.

Functionalism

The second major theory of the mind, functionalism, defines types of mental states in terms of their causal roles relative to sensory stimulation, other mental states, and physical states or behaviour. Pain, for example, might be defined as the type of neurophysiological state that is caused by things like cuts and burns and that causes mental states such as fear and "pain behaviour" such as saying "ouch". Functionalism avoids the second objection against the type-type identity theory mentioned above – that it seems possible for two people to be in the same mental state but not in the same neurophysiological state

– because it is not committed to the idea that the neurophysiological state that plays the causal role of pain must be the same in all people, or the same in people as in non-human creatures. This point was often expressed by saying that functional states exhibit "multiple realizability".

Functionalism was inspired in part by the development of the computer, which was understood in terms of the distinction between hardware, or the physical machine, and software, or the function that the computer performs. It also was influenced by the earlier idea of a Turing machine, named after the English mathematician Alan Turing (see Chapter 7). A Turing machine is an abstract device that receives information as input and produces other information as output, the particular output depending on the input, the internal state of the machine, and a finite set of rules that associate input and machine-state with output. Turing defined intelligence functionally, in the sense that for him anything that possessed the ability to transform information from one form into another, as the Turing machine does, counted as intelligent to some degree. This understanding of intelligence was the basis of what came to be known as the Turing test, which proposed that, if a computer could answer questions posed by a remote human interrogator in such a way that the interrogator could not distinguish the computer's answers from those of a human subject, then the computer could be said to be intelligent and to think. Following Turing, the American philosopher Hilary Putnam held that the human brain is basically a sophisticated Turing machine, and his functionalism was accordingly called "Turing machine functionalism". This became the basis of the later theory known as strong artificial intelligence (or strong AI), which asserts that the brain is a kind of computer and the mind a kind of computer program.

In the 1980s the American philosopher John Searle mounted a challenge to strong AI. Searle's objections were based on the observation that the operation of a computer program consists of the manipulation of certain symbols according to rules that refer only to the symbols' formal or syntactic properties and not to their semantic ones. In his so-called "Chinese-room argument", Searle attempted to show that there is more to thinking than this kind of rule-governed manipulation of symbols. The argument involves a situation in which a person who does not understand Chinese is locked in a room. They are handed written questions in Chinese, to which they must provide written Chinese answers. With the aid of a computer program or a rule book that matches questions in Chinese with appropriate Chinese answers, the person could simulate the behaviour of a person who understands Chinese. Thus, a Turing test would count such a person as understanding Chinese. But by hypothesis, they do not have that understanding. Hence, understanding Chinese does not consist merely in the ability to manipulate Chinese symbols. What the functionalist theory leaves out and cannot account for, according to Searle, are the semantic properties of the Chinese symbols, which are what the Chinese speaker understands. In a similar way, the Turing-functionalist definition of thinking as the manipulation of symbols according to syntactic rules is deficient because it leaves out the symbols' semantic properties.

A more general objection to functionalism involves what is called the "inverted spectrum". It is entirely conceivable, according to this objection, that two humans could possess inverted colour spectra without knowing it. The two may use the word red, for example, in exactly the same way, and yet the colour sensations they experience when they see red things may be different. Because the sensations of the two people play the same causal role for each, however, functionalism is committed to the claim that the sensations are the same.

Counter-examples such as these demonstrated that similarity of function does not guarantee identity of subjective experience, and accordingly that functionalism fails as an analysis of mental content. Putnam eventually agreed with these and other criticisms, and in the 1990s he abandoned the view he had created.

Eliminative Materialism

The most radical theory of the mind developed in this period is eliminative materialism. Introduced in the late 1980s and refined and modified throughout the 1990s, it contended that scientific theory does not require reference to the mental states posited in informal, or "folk", psychology, such as thoughts, beliefs, desires, and intentions. The correct view of the human mind, according to eliminative materialism, is that there are no mental states in the folk-psychological sense and that the mind is nothing more or less than the brain. Furthermore, because there are no mental states, both the identity theory and functionalism are trying to do the impossible, i.e. to reduce non-existent mental events to neural activity. Just as late eighteenth-century chemical theory did not try to reduce the fictional concept of phlogiston to molecular states but simply dispensed with any reference to it, so the entire mentalistic vocabulary of folk psychology can be eliminated in a sophisticated scientific theory of the mind. Such a theory will simply describe how the brain works.

Three main objections were raised against this view. The first was that it failed to explain how semantic properties such as meaning, truth, and reference could be elicited from, or instantiated in, neural activity. In brief, this objection argued that it is simply a conceptual mistake to try to ascribe truth or falsity, or any semantic property, to brain processes, as eliminative materialism would seem to require. The second objection was that eliminative materialism denied the existence of certain things that all accept as real: namely, felt sensations (known as "qualia"). To deny that qualia exist is tantamount to saying that there are no such things as sounds, only air vibrating at various frequencies.

The third objection to eliminative materialism emphasized the fact that each person has access to their own mental experiences in a way that no other person has. Pains and visual images, as well as countless other kinds of thought, possess a kind of subjectivity that cannot be captured in a purely scientific account, because scientific descriptions concern only the objective properties of natural phenomena. There were many variants of this position. Among the philosophers and scientists who rejected reductivism on these or other grounds were Searle, Roderick Chisholm, Zeno Vendler, Thomas Nagel, Roger Penrose, Alastair Hannay, and J.R. Smythies.

Consciousness

In the early 19th century the concept of consciousness was variously considered. Some philosophers regarded it as a kind of substance, or "mental stuff", quite different from the material substance of the physical world. Others thought of it as an attribute characterized by sensation and voluntary movement, which separated animals and humans from lower forms of life and also described the difference between the normal waking state of animals and humans and their condition when asleep, in a coma, or under anaesthesia (the latter condition was described as unconsciousness). Other descriptions included an analysis of consciousness as a form of relationship or act of the mind toward objects in nature, and a view that consciousness was a continuous field or stream of essentially mental "sense data", roughly similar to the "ideas" of earlier empirical philosophers.

The method employed by most early writers in observing consciousness was introspection – looking within one's own mind to discover the laws of its operation. The limitations of the method became apparent when it was found that because of differing preconceptions, trained observers in the laboratory often could not agree on fundamental observations.

The Behaviourist View

The failure of introspection to reveal consistent laws led to the rejection of all mental states as proper subjects of scientific study. In behaviourist psychology, derived primarily from work by the American psychologist John B. Watson in the early 1900s, the concept of consciousness was irrelevant to the objective investigation of human behaviour and was doctrinally ignored in research. Neobehaviourists, however, adopted a more liberal posture toward mentalistic states such as consciousness.

Neurophysiological Mechanisms

That consciousness depends on the function of the brain has been known from ancient times. Although detailed understanding of the neural mechanisms of consciousness has not been achieved, correlations between states of consciousness and functions of the brain are possible. Levels of consciousness in terms of levels of alertness or responsiveness are correlated with patterns of electrical activity of the brain (brain waves) recorded by an electroencephalograph. During wide-awake consciousness the pattern of brain waves consists of rapid irregular waves of low amplitude or voltage. In contrast, during sleep, when consciousness can be said to be minimal, the brain waves are much slower and of greater amplitude, often coming in periodic bursts of slow waxing and waning amplitude.

Both behavioural levels of consciousness and the correlated patterns of electrical activity are related to the function of a part of the brainstem called the reticular formation. Electrical stimulation of the ascending reticular systems arouses a sleeping cat to alert consciousness and simultaneously activates its brain waves to the waking pattern.

It was once supposed that the neurophysiological mechanisms subserving consciousness and the higher mental processes must reside in the cortex. It is more likely, however, that the cortex serves the more specialized functions of integrating patterns of sensory experience and organizing motor patterns and that the ascending reticular system represents the neural structures most critically related to consciousness. The brainstem reticular formation should not, however, be called the seat of consciousness. It represents an integrative focus, functioning through its widespread interconnections with the cortex and other regions of the brain.

Executives, Buffers, and HOTs

Since the 1980s various theories have been developed to explain the role that consciousness plays in the mental lives of human beings and other animals. "Executive" theories stress the role of conscious states in deliberation and planning. Many researchers, however, doubt that all such executive activities are conscious, and some suspect that conscious states may play a more tangential role in determining action. Some have even proposed that conscious states are mostly a kind of "afterthought", reflecting executive decisions that have already been made at an unconscious level.

According to "buffer" theories, consciousness consists of a specific relation between a subject and a specific location in the brain in which material is stored for certain purposes, such as introspection. The American linguist Ray Jackendoff has made the further interesting suggestion that such material is confined to relatively low-level sensory information.

An important family of much more specific theories are variants of the idea that consciousness involves some kind of state directed at another state. One proposal is that consciousness involves some kind of "internal scanning" or "perception"; another is that it involves an explicit "higher-order thought" (HOT), i.e. a thought that one is in a specific mental state. According to the latter view, a person's thought that they desire to cross the road is conscious only if they also have the belief that they desire to cross the road. (This does not imply, of course, that HOTs themselves are always conscious.) Some philosophers have argued that the HOT must actually be occurring at the time the target-thought is conscious; others have held that the subject must simply be disposed to have the relevant HOT.

Hallucination

A historical survey of the study of hallucinations reflects the development of scientific thought in psychiatry, psychology, and neurobiology. By 1838 the significant relationship between the content of dreams and of hallucinations had been pointed out. In the 1840s the occurrence of hallucinations under a wide variety of conditions (including psychological and physical stress), as well as their genesis through the effects of such drugs as stramonium and hashish, had been described.

In 1845, French physician Alexandre-Jacques-François Brierre de Boismont described many instances of hallucinations associated with intense concentration, or with musing, or simply occurring in the course of psychiatric disorder. In the last half of the nineteenth century, studies of hallucinations continued. Investigators in France were particularly oriented toward abnormal psychological function, and from this came descriptions of hallucinosis during sleepwalking and related reactions. In the 1880s English neurologist John Hughlings Jackson described hallucination as being released or triggered by the nervous system.

Other definitions of the term emerged later. Swiss psychiatrist Eugen Bleuler (1857–1939) defined hallucinations as "perceptions without corresponding stimuli from without", while the *Psychiatric Dictionary* in 1940 referred to hallucination as the "apparent perception of an external object when no such object is present". A spirited interest in hallucinations continued well into the twentieth century. Sigmund Freud's concepts of conscious and unconscious activities added new significance to the content of dreams and hallucinations. It was theorized that infants normally hallucinate the objects and processes that give them gratification. Although the notion has since been disputed, this "regression" hypothesis (i.e. that hallucinating is a regression, or return, to infantile ways) is still employed, especially by those who find it clinically useful. During the same period, others put forth theories that were more broadly biological than Freud's but that had more points in common with Freud than with each other.

The general theory of hallucinations rests upon two fundamental assumptions. One assumption states that life experiences influence the brain in such a way as to leave, in the brain, enduring physical changes that have variously been called neural traces, templates, or engrams. Ideas and images are held to derive from the incorporation and activation of these engrams in complex circuits involving nerve cells. Such circuits in the cortex (outer layers) of the brain appear to subserve the neurophysiology of memory, thought, imagination, and fantasy. The emotions associated with these intellectual and perceptual functions seem to be mediated through cortex connections with the deeper parts of the brain (the limbic system or "visceral brain", for example), thus permitting a dynamic interplay between perception and emotion through transactions that appear to take place largely at unconscious levels.

Conscious awareness is found to be mediated by the ascending midbrain reticular activating system (a network of nerve cells in the brainstem). Analyses of hallucinations reported by sufferers of neurological disorders and by neurosurgical patients in whom the brain is stimulated electrically have shown the importance of the temporal lobes (at the sides of the brain) to auditory hallucinations, for example, and of other functionally relevant parts of the brain in this process.

A second assumption states that the total human personality is best understood in terms of the constant interplay of forces that continually emanate from inside (as internal physiological activity) and from outside the individual (as sensory stimuli). Such transactions between the environment and the individual may be said to exert an integrating and organizing influence upon memory traces stored in the nervous system and to affect the patterns in which sensory engrams are activated to produce experiences called images, fantasies, dreams, or hallucinations, as well as the emotions associated with these patterns. If such a constantly shifting balance exists between internal and external environmental forces, physiological considerations (e.g. brain function) as well as cultural and experiential factors emerge as major determinants of the content and meaning of hallucinations.

The brain is bombarded constantly by sensory impulses, but most of these are excluded from consciousness in a dynamically shifting, selective fashion. The exclusion seems to be accomplished through the exercise of integrative inner mechanisms that focus one's awareness on selected parts of potential experience. (The sound of a ticking clock, for example, fades in and out of awareness.) Functioning simultaneously, these mechanisms survey information that is stored within the brain, select tiny samples needed to give adaptive significance to the incoming flow of information, and bring forth only a few items for actual recall from the brain's extensive "memory banks".

PART 3

WHAT HAPPENS WHEN THINGS GO WRONG?

10

ILLNESS AND THERAPY

Throughout a person's lifetime, billions of nerve cells are produced, grow, and organize themselves into effective, functionally active systems that ordinarily remain in working order. The motivation of neuroscientists who research this most complex of all machines is two-fold: to understand human behaviour better, and to discover ways to prevent or cure many devastating brain disorders. There are more than 1,000 disorders of the brain and nervous system, some of which will be discussed below. These diseases result in more hospitalizations than any other disease group, including heart disease and cancer. In addition to neurological disease and damage, there are mental disorders, which will be considered in the next chapter.

Some memory failure is almost universal during old age, particularly in forgetfulness for names and in the reduced ability to learn. Many people of advanced age, nevertheless, show adequate memory function if they suffer no brain disease. Impairment of memory is a characteristic early sign of senility, as well as of hardening of the brain arteries (cerebral arteriosclerosis) at any age, with exaggerated forgetfulness for recent events and progressive failure in memory for experiences that preceded the disorder. As arteriosclerotic brain disease progresses, amnesia tends to extend further into the past, embracing personal experience and general or common information. When the symptoms are almost those of Korsakoff's syndrome, the disturbance is called presbyophrenia. In most cases the amnesia is complicated by failure in judgement and changes in character. It has been suggested that severe memory defect in an elderly person carries a poor prognosis, being related to such factors as a shortened survival time and an increased death rate.

A Swiss psychiatrist, Eugen Bleuler, held that amnesia results only from a diffuse disorder of the outer layers (cortex) of the brain and suggested that memory depends on the integrity of the cortex as a whole. Indeed, the removal of brain tissue from rats and monkeys in experimental studies has indicated that retention of complex habits by the animals depends on the total amount of cortex that remains. It was claimed that the degree to which memory is lost depends not on where the brain is injured but on the extent of the damage. (This is the "law" of mass action, which asserts that the brain functions in a unitary manner, i.e. as a whole.) While the extent of diffuse brain damage is roughly related to the severity of memory defect, the principle of mass action is manifestly inadequate. Whatever its physical basis, memory seems to depend on the integrity of relatively limited parts of the brain, rather than on that organ (or even the cortex) as a whole.

Severe and highly specific amnesic symptoms principally stem from damage to such brain structures as the mammillary bodies, circumscribed parts of the thalamus, and of the temporal lobe (e.g. the hippocampus). While the ability to store new experience (and perhaps to retrieve well-established memories) appears to depend on a distinct neural system involving the temporal cortex and limited parts of the thalamus and hypothalamus, understanding of the neuroanatomy of memory remains sketchy enough to generate major differences of opinion. French and German workers tend to stress the role of the mammillary bodies, while American investigators tend to implicate the thalamus. It has been pointed out that circumscribed damage to the mammillary bodies is not invariably associated with memory defect; cases of amnesia evidently occur in which these structures are spared. Nevertheless, implication of the mammillary bodies in a large number of verified cases of Korsakoff's syndrome seems incontrovertible. Injury to other neural tissues (e.g. the so-called fornix bundle deep within the brain) that anatomically might be expected to produce severe memory disorder rarely does so. While evidence for amnesia as a sign of localized brain damage is impressive, much remains to be understood about the physical system that sustains memory.

Neurological Disease

Huntington's Disease

Huntington's disease, also called Huntington's chorea, is a relatively rare, and invariably fatal, hereditary neurological disease that is characterized by irregular and involuntary movements of the muscles and progressive loss of cognitive ability. The disease was first described by the American physician George Huntington in 1872.

Symptoms of Huntington's disease usually appear between the ages of 35 and 50 and worsen over time. They begin with occasional jerking or writhing movements, called choreiform movements, or what appear to be minor problems with coordination; these movements, which are absent during sleep, worsen over the next few years and progress to random, uncontrollable, and often violent twitchings and jerks. Symptoms of mental deterioration may appear including apathy, fatigue, irritability, restlessness, or moodiness; these symptoms may progress to memory loss, dementia, bipolar disorder, or schizophrenia.

A child of someone with Huntington's disease has a 50 per cent chance of inheriting the genetic mutation associated with the disease, and all individuals who inherit the mutation will eventually develop the disease. The genetic mutation that causes Huntington's disease occurs in a gene known as *HD* (officially named huntingtin [Huntington's disease]). This gene, which is located on human chromosome 4, encodes a protein called huntingtin, which is distributed in certain regions of the brain, as well as other tissues of the body. Mutated forms of the *HD* gene contain abnormally repeated segments of deoxyribonucleic acid (DNA) called CAG trinucleotide repeats. These repeated segments result in the synthesis of huntingtin proteins that contain long stretches of molecules of the amino acid glutamine. When these abnormal huntingtin proteins are cut into fragments during processing by cellular enzymes, molecules of glutamine project out from the ends of the protein fragments, causing the fragments to adhere to other proteins. The resulting clumps of proteins have the potential to cause neuron (nerve cell) dysfunction. The formation of abnormal huntingtin proteins leads to the degeneration and eventual death of neurons in the basal ganglia, a pair of nerve clusters deep within the brain that control movement.

The progression and severity of Huntington's disease are associated with the length of the CAG trinucleotide repeat region in the *HD* gene. For example, the trinucleotide region in *HD* appears to expand during middle age, coinciding with the onset of symptoms, and may also expand from one generation to the next, causing a form of the disease known as anticipation, in which symptoms develop at an earlier age in the offspring of affected individuals.

While a genetic test is available for Huntington's disease, no effective therapy or cure is available for the disorder, although choreiform movements may be partially and temporarily suppressed by phenothiazines or other antipsychotic medications.

Parkinson's Disease

Parkinsonism is a chronic neurological disorder characterized by a progressive loss of motor function resulting from the degeneration of neurons in the area of the brain that controls voluntary movement.

Parkinsonism was first described in 1817 by the British physician James Parkinson in his *Essay on the Shaking Palsy*. Various types of the disorder are now recognized, but the disease described by Parkinson, called Parkinson's disease, is the most common form. Parkinson's disease is also called primary parkinsonism, paralysis agitans, or idiopathic parkinsonism. The onset of Parkinson's disease typically occurs at about 60 years of age. It is rarely inherited. Parkinson's disease often begins with a slight tremor of the thumb and forefinger, sometimes called "pill-rolling", and slowly progresses over 10 to 20 years, resulting in paralysis, dementia, and death.

The four main signs of parkinsonism are tremors of resting muscles, particularly of the hands; muscular rigidity of the arms, legs, and neck; difficulty in initiating movement (bradykinesia); and loss of balance. A variety of other features may accompany these characteristics, including a lack of facial expression (known as "masked face"), difficulty in swallowing or speaking, stooped posture, a shuffling gait, depression, and dementia.

Parkinsonism results from the deterioration of neurons in the region of the brain called the substantia nigra. These neurons normally produce the neurotransmitter dopamine, which sends signals to the basal ganglia, a mass of nerve fibres that helps to initiate and control patterns of movement. Dopamine functions in the brain as an inhibitor of nerve impulses and is involved in suppressing unintended movement. When dopamine-producing (dopaminergic) neurons are damaged or destroyed, dopamine levels drop and the normal signalling system is disrupted. The features of parkinsonism do not appear until 60 to 80 per cent of these neurons are destroyed.

Although the cause of neuronal deterioration in primary parkinsonism is unknown, causal agents have been identified for some types of the disorder, referred to as secondary parkinsonism. A viral infection of the brain that caused a worldwide pandemic of encephalitis lethargica (sleeping sickness) just after the First World War resulted in the development of post-encephalitic parkinsonism in some survivors. Toxin-induced parkinsonism is caused by carbon monoxide, manganese, or cyanide poisoning. A neurotoxin known as MPTP (1-methyl-4-phenyl-1,2,3,6-tetrahydropyridine), previously found in contaminated heroin, also causes a form of toxin-induced parkinsonism. The ability of this substance to destroy neurons suggests that an environmental toxin similar to MPTP may be responsible for Parkinson's disease. Pugilistic parkinsonism results from head trauma and has affected professional boxers such as Jack Dempsey and Muhammad Ali. The parkinsonism-dementia complex of Guam, which occurs among the Chamorro people of the Pacific Mariana Islands, is also thought to result from an unidentified environmental agent. In some individuals a genetic defect is thought to incur susceptibility to the disease. Parkinsonism-plus disease, or multiple-system degenerations, includes diseases in which the main features of parkinsonism are accompanied by other symptoms. Parkinsonism may appear in patients with other neurological disorders such as Huntington's disease, Alzheimer's disease, and Creutzfeldt–Jakob disease.

Both medical and surgical therapies are used to treat parkinsonism. The medication levodopa (L-dopa, Larodopa), a precursor of dopamine, is used in conjunction with the medication carbidopa to alleviate symptoms, although this treatment tends to become less effective over time. Other medications used are selegiline, a type of drug that slows the breakdown of dopamine, and bromocriptine and pergolide, two drugs that mimic the effects of dopamine. Surgical procedures are used to treat parkinsonism patients who have failed to respond to medications. Pallidotomy involves destroying a part of the brain structure called the globus pallidus that is involved in motor control. Pallidotomy may improve symptoms such as tremors, rigidity, and bradykinesia. Cryothalamotomy destroys the area of the brain that produces tremors by inserting a probe into the thalamus. Restorative surgery is an experimental technique that replaces the lost dopaminergic neurons of the patient with dopamine-producing fetal brain tissue.

Brain Damage

The frontal lobes are the part of the brain most remote from sensory input and whose functions are the most difficult to capture. They can be thought of as the executive that controls and directs the operation of brain systems dealing with cognitive function. The deficits seen after frontal lobe damage are described as a "dysexecutive syndrome".

Frontal lobe damage can affect people in any of several ways. On the one hand, they may have difficulty initiating a task or a behaviour, in extreme cases being virtually unable to move or speak, but more often they will simply have difficulty in initiating a task. On the other hand, individuals with frontal lobe damage may perservate, being apparently unable to stop a behaviour once it is started. Rather than appearing apathetic and hypoactive, patients may be uninhibited and may appear rude. Such people may also have difficulty in planning and problem solving and may be incapable of creative thinking. Mild cases of this deficit may be determined by a difficulty in solving mental arithmetic problems that are filled with words, even though the patient is capable of remembering the question and performing the required calculation. In such cases it appears that the patient simply cannot select the appropriate cognitive strategy to solve the problem.

A unifying theme in these disorders is the notion of inadequate control of organization of pieces of behaviour that may in themselves be well formed. Patients with frontal lobe damage are easily distracted. Although their deficits may be superficially less dramatic than those associated with posterior lesions, they can have a drastic effect on everyday function. Irritability and personality change are also frequently seen after frontal lobe damage.

Craniocerebral Trauma

The concussive and shearing stresses of head injury may cause concussion, contusion of the brain (most often of the tips of the frontal and temporal lobes, called contrecoup injury), or laceration of the brain tissue. In the last two cases, neurological deficits are detected at the time of injury, and with laceration (as in a depressed fracture of the skull) or bleeding into the brain, post-traumatic epilepsy is possible.

Extradural hematomas, often from tearing of the middle meningeal artery, may result as a complication of a head injury. Arterial blood, pumped into the space between the dura and the inside of the skull, compresses the brain downward through the tentorium or the foramen magnum. Surgical removal of the clot is necessary. Subdural hematomas usually develop more slowly and may sometimes take weeks to form; they follow the rupture of small veins bridging the gap between the surface of the brain and the meninges. Headache, seizures, intellectual decline, and symptoms similar to those of extradural hematomas may occur. Removal of the clot is the usual treatment.

Other complications of head injury include cranial nerve palsies, subarachnoid haemorrhage, thrombosis of a carotid artery, focal deficits, and cerebrospinal fluid leakage, which may lead to intracranial infection. Later consequences include dementia, seizures, irritability, fatigue, headaches, insomnia, loss of concentration, poor memory, and loss of energy. Repeated minor head injuries, which may occur in some boxers, may also lead to dementia and to a Parkinson-like syndrome.

Disease of the Cerebral Hemispheres

The frontal lobe is involved with many of the components of intelligence (foresight, planning, and comprehension), with mood, with motor activity on the opposite side of the body, and (in the case of the dominant hemisphere) with speech production. Swelling of the underside of the frontal lobe may compress the first cranial nerve and result in the loss of smell. Irritation of the frontal cortex may also cause either generalized or local motor epileptic seizures, the latter involving the opposite side of the body.

Damage to the dominant temporal lobe, located inferior to the lateral sulcus, results in difficulty with comprehension of spoken speech. The right temporal lobe (usually non-dominant for speech) has a special role in the appreciation of non-language sounds such as music. Irritation of a temporal lobe may lead to auditory or olfactory hallucinations. Memory functions are duplicated in the two temporal lobes; if one lobe is damaged, there may be little effect, but bilateral damage leads to a permanent inability to learn new data.

In most people the left parietal lobe shares control of the comprehension of spoken and written language and of arithmetic, interprets the difference between right and left, identifies body parts, and determines how to perform meaningful motor actions. Damage to this lobe, located posterior to the central sulcus, leads to forms of apraxia, the inability to perform purposeful actions. The right parietal lobe is concerned with visuospatial orientation, and damage typically leads to deficits such as dressing apraxia (inability to put on clothes), constructional apraxia (difficulty in creating or copying two- or three-dimensional forms), and sensory competition, or sensory extinction, which is an inability to recognize two stimuli when both are presented together on opposite sides of the body – most easily demonstrated in the sensations of touch and vision. Each parietal lobe is also involved with so-called cortical sensation or discriminative touch, the analysis and interpretation of touch sensations originating on the other side of the body. Damage to the parietal lobe can cause a form of agnosia in which sensation is present but interpretation or comprehension is lacking. Irritation of the parietal lobe also leads to tactile hallucinations, the false perception of touch sensations on the other side of the body.

The occipital lobes, which lie below and behind the parieto-occipital sulcus, are almost exclusively involved with the reception of visual impulses. Damage to one side results in homonymous hemianopia, the loss of all sight in the field of vision on the opposite side. Compression of the optic chiasm, usually by a tumour of the pituitary fossa, may result in the "blinkers" effect. At the optic chiasm the optic nerve fibres from the nasal halves of the right and left retinas cross to the opposite side. Because the nasal retinas "see" the temporal fields (the right nasal retina receiving impulses from objects to the right, the left from objects to the left), a patient with a lesion of the optic chiasm is able to see straight ahead but not to either side. This is called bitemporal hemianopia.

Irritation of the occipital lobe causes the subject to see hallucinations. If the lesion is far back in the lobe, the hallucinations may be of unformed lights, colours, or shapes. However, they also may be vivid and sharply defined pictures, as though a videotape of previous visual experiences were being replayed, if the lesion is farther forward and in the area where the parietal, temporal, and occipital lobes adjoin. This area of the cortex appears to be involved with the analysis and storage of complex perceptions.

Disease of the Cerebrum

Cerebral Palsy

The term cerebral palsy encompasses all of the conditions that damage the brain around or before the time of birth – other than developmental causes. Approximately six cases of cerebral palsy occur in every 1000 live births in the Western world. Hypoxia and asphyxia during a prolonged and difficult labour are the most common causes, but improvements in obstetrical care have reduced the incidence of the condition. Cerebral palsy is characterized by a delay in motor development, spasticity, weakness of the limbs, athetosis, or ataxia. Sensory, visual, and cognitive defects may be detected later in life. Intellectual disability occurs in about half of children with the condition.

There are four types of cerebral palsy: spastic, athetoid, ataxic, and mixed. In the spastic type, there is a severe paralysis of voluntary movements, with spastic contractions of the extremities either on one side of the body (hemiplegia) or on both sides (diplegia). In spastic diplegia, spastic contractions and paralysis are usually more prominent in the lower extremities than in the arms and hands (Little diplegia), or only the legs may be affected (paraplegia). The cerebral damage causing spastic cerebral palsy primarily affects the neurons and connections of the cerebral cortex, either of one cerebral hemisphere (contralateral to paralysis), as in infantile hemiplegia, or of both hemispheres, as in diplegia.

In the athetoid type of cerebral palsy, paralysis of voluntary movements may not occur, and spastic contractions may be slight or absent. Instead, there are slow, involuntary spasms of the face, neck, and extremities, either on one side (hemiathetosis) or, more frequently, on both sides (double athetosis), with resulting involuntary movements in the whole body or its parts, facial grimacing, and inarticulate speech (dysarthria) – all of which increase under stress or excitement. Damage to the brain particularly affects the basal ganglia underlying the cerebral cortex.

Ataxic cerebral palsy is a rare form of the condition that is characterized by poor coordination, muscle weakness, an unsteady gait, and difficulty performing rapid or fine movements. If symptoms of two or more types are present, most often spastic and athetoid, an individual is diagnosed with mixed cerebral palsy.

Cerebral palsy does not necessarily include intellectual disability; many children affected with cerebral palsy are mentally competent. However, any cerebral disorder in early life may result in impairment, sometimes severe, of intellectual and emotional development. Epileptic attacks in the form of convulsive seizures, especially in the parts of the body affected by paralysis, occur in many children with cerebral palsy. In the spastic type of cerebral palsy, intellectual disability and epileptic attacks are particularly frequent. In the athetoid type, the incidence of severe intellectual disability is much lower, and occurrence of convulsive seizures is rare. Children affected with athetoid cerebral palsy may be perceptive and intelligent; however, because of the involuntary movements and dysarthria, they are often unable to communicate by intelligible words or signs.

The causes of cerebral palsy are multiple but basically involve a malfunctioning of the complex neuronal circuits of the basal ganglia and the cerebral cortex. Heredity plays only a small role. It may manifest itself in malformations of neurons, interstitial tissues, or blood vessels of the brain that may produce tumours, or it may express itself in an abnormal chemistry of the brain. More common causes of the condition are fetal diseases and embryonic malformations of the brain. Incompatibility of blood types of the mother and fetus, leading to severe jaundice at birth, may cause brain damage and cerebral palsy. Respiratory problems of the fetus during birth may indicate earlier brain damage. Paediatric infections, severe head injuries, and poisoning are other less common causes of cerebral palsy.

There is no cure for cerebral palsy; treatment includes medications that relax the muscles and prevent seizures. The basic programme of treatment aims at the psychological management, education, and training of the child to develop sensory, motor, and intellectual assets, in order to compensate for the physical liabilities of the disorder.

Dementia

Dementia is the chronic, usually progressive deterioration of intellectual capacity associated with the widespread loss of nerve cells and the shrinkage of brain tissue. It is most commonly seen in the elderly (senile dementia), though it is not part of the normal ageing process and can affect people of any age.

The most common irreversible dementia is Alzheimer's disease. This condition begins with memory loss, which may first appear to be simple absent-mindedness or forgetfulness. As dementia progresses, the loss of memory broadens in scope until the individual can no longer remember basic social and survival skills or function independently. Language, spatial or temporal orientation, judgement, or other cognitive capacities may decline, and personality changes may also occur. Dementia is also present in other degenerative brain diseases including Pick disease and Parkinson's disease.

The second most common cause of dementia is hypertension (high blood pressure) or other vascular conditions. This type of dementia, called multi-infarct, or vascular, dementia results from a series of small strokes that progressively destroy the brain. Dementia can also be caused by Huntington's disease, syphilis, multiple sclerosis, acquired immune deficiency syndrome (AIDS), and some types of encephalitis. Treatable dementias occur in hypothyroidism, other metabolic diseases, and some malignant tumours. Treatment of the underlying disease in these cases may inhibit the progress of dementia but usually does not reverse it.

Alzheimer's Disease

Alzheimer's disease is a degenerative brain disorder that develops in mid- to late adulthood. It results in a progressive and irreversible decline in memory and a deterioration of various other cognitive abilities. The disease is characterized by the destruction of nerve cells and neural connections in the cerebral cortex of the brain and by a significant loss of brain mass. The disease was first described in 1906 by Alois Alzheimer, a German neuropathologist.

Alzheimer's disease is the most common form of dementia. The disease develops differently among individuals, suggesting that more than one pathological process may lead to the same outcome. Typically, the first symptom to appear is forgetfulness. As the disease progresses, memory loss becomes more severe, and language, perceptual, and motor skills deteriorate. Mood becomes unstable, and the individual tends to become irritable and more sensitive to stress and may become intermittently angry, anxious, or depressed. In advanced stages, the individual becomes unresponsive and loses mobility and control of body functions; death ensues after a disease course lasting from two to 20 years.

About 10 per cent of those who develop the disease are younger than 60 years of age. These cases, referred to as early-onset familial Alzheimer's disease, result from an inherited genetic mutation. The majority of cases of Alzheimer's disease, however, develop after age 60 (late-onset); they usually occur sporadically, i.e. in individuals with no family history of the disease, although a genetic factor has been identified that is thought to predispose these individuals to the disorder.

Other features have been noted in the brains of many people with Alzheimer's disease. One feature is a deficiency of the neurotransmitter acetylcholine; neurons containing acetylcholine play an important role in memory. A second feature is abnormal insulin signalling in the brain. Under normal conditions, insulin binds to insulin receptors, which are expressed in great numbers on the membranes of neurons, to facilitate neuronal uptake of glucose, which the brain depends upon to carry out its many functions.

However, neurons in the brains of patients with Alzheimer's disease have very few, if any, insulin receptors and therefore are resistant to the actions of insulin. As a result of the inability of insulin to bind to the neurons, it accumulates in the blood serum, leading to a condition known as hyperinsulinemia (abnormally high serum levels of insulin). Hyperinsulinemia in the brain is suspected to stimulate inflammation that in turn stimulates the formation of neuritic plaques. Abnormal insulin signalling in the brain has also been associated with nerve cell dysfunction and death, decreased levels of acetylcholine, and decreased levels of transthyretin, a protein that normally binds to and transports beta-amyloid proteins out of the brain.

Underlying genetic defects have been identified for both late- and early-onset cases of Alzheimer's disease. A defect in the gene that codes for amyloid precursor protein may increase the production or deposition of beta-amyloid, which forms the core of neuritic plaques. This gene, however, is responsible for only two to three per cent of all early-onset cases of the disease; the remainder are attributed to two other genes.

There is no cure for Alzheimer's disease. The medication tacrine slightly slows the progression of the disease by slowing the breakdown of acetylcholine, but it is not effective in all patients and can become toxic to the liver. Most treatment aims to control the depression, behavioural problems, and insomnia that often accompany the disease.

Epilepsy

Epilepsy is characterized by sudden and recurrent seizures caused by excessive signalling of nerve cells in the brain. Seizures may include convulsions, momentary lapses of consciousness, strange movements or sensations in parts of the body, odd behaviours, and emotional disturbances. Epileptic seizures typically last one to two minutes but can be followed by weakness, confusion, or unresponsiveness. Epilepsy is a relatively common disorder affecting about 40 to 50 million people worldwide; it is slightly more common in males than females. Causes of the disorder include brain defects, head trauma, infectious diseases, stroke, brain tumours, or genetic or developmental abnormalities. Several types of epileptic disorders are hereditary. Cysticercosis, a parasitic infection of the brain, is a common cause of epilepsy in the developing world. About half of epileptic seizures have an unknown cause and are called idiopathic.

A partial seizure originates in a specific area of the brain. Partial seizures consist of abnormal sensations or movements, and a lapse of consciousness may occur. Epileptic individuals with partial seizures may experience unusual sensations called auras that precede the onset of a seizure. Auras may include unpleasant odours or tastes, the sensation that unfamiliar surroundings seem familiar (déjà vu), and visual or auditory hallucinations that last from a fraction of a second to a few seconds. The individual may also experience intense fear, abdominal pain or discomfort, or an awareness of increased respiration rate or heartbeat. The form of the onset of a seizure is, in most cases, the same from attack to attack. After experiencing the aura, the individual becomes unresponsive but may examine objects closely or walk around.

Jacksonian seizures are partial seizures that begin in one part of the body such as the side of the face, the toes on one foot, or the fingers on one hand. The jerking movements then spread to other muscles on the same side of the body. This type of seizure is associated with a lesion or defect in the area of the cerebral cortex that controls voluntary movement.

Complex partial seizures, also called psychomotor seizures, are characterized by a clouding of consciousness and by strange, repetitious movements called automatisms. On recovery from the seizure, which usually lasts from one to three minutes, the individual has no memory of the attack, except for the aura. Occasionally, frequent mild complex partial seizures may merge into a prolonged period of confusion, which can last for hours or days with fluctuating levels of awareness and strange behaviour. Complex partial attacks may be caused by lesions in the frontal lobe or the temporal lobe.

Generalized seizures are the result of abnormal electrical activity in most or all of the brain. This type of seizure is characterized by convulsions, short absences of consciousness, generalized muscle jerks (clonic seizures), and loss of muscle tone (tonic seizures), with falling.

Generalized tonic-clonic seizures, sometimes referred to by the older term grand mal, are commonly known as convulsions. A person undergoing a convulsion loses consciousness and falls to the ground. The fall is sometimes preceded by a shrill scream caused by forcible expiration of air as the respiratory and laryngeal muscles suddenly contract. After the fall, the body stiffens because of generalized tonic contraction of the muscles; the lower limbs are usually extended and the upper limbs flexed. During the tonic phase, which lasts less than a minute, respiration stops because of sustained contraction of the respiratory muscles. Following the tonic stage, clonic (jerking) movements occur in the arms and legs.

The tongue may be bitten during involuntary contraction of the jaw muscles, and urinary incontinence may occur. Usually, the entire generalized tonic-clonic seizure is over in less than five minutes. Immediately afterward, the individual is usually confused and sleepy and may have a headache but will not remember the seizure.

Primary generalized, or absence, epilepsy is characterized by repeated lapses of consciousness that generally last less than 15 seconds each and usually occur many times a day. This type of seizure is sometimes referred to by the older term petit mal. Minor movements such as blinking may be associated with absence seizures. After the short interruption of consciousness, the individual is mentally clear and able to resume previous activity. Absence seizures occur mainly in children and do not appear initially after age 20; they tend to disappear before or during early adulthood. At times absence seizures can be nearly continuous, and the individual may appear to be in a clouded, partially responsive state for minutes or hours.

A person with recurrent seizures is diagnosed with epilepsy. A complete physical examination, blood tests, and a neurological evaluation may be necessary to identify the cause of the disorder. EEG monitoring is performed to detect abnormalities in the electrical activity of the brain. MRI, PET, SPECT, or MRS (see Chapter 3 for imaging techniques) may be used to locate structural or biochemical brain abnormalities.

Treatment

Most people with epilepsy have seizures that can be controlled with anti-epileptic medications such as valproate, ethosuximide, clonazepam, carbamazepine, and primidone; these medications decrease the amount of neuronal activity in the brain.

Brain damage caused by epilepsy usually cannot be reversed. Epileptic seizures that cannot be treated with medication may be reduced by surgery that removes the epileptogenic area of the brain. Other treatment strategies include vagus nerve stimulation, a diet high in fat and low in carbohydrates (ketogenic diet), and behavioural therapy. It may be necessary for epileptic individuals to refrain from driving, operating hazardous machinery, or swimming because of the temporary loss of control that occurs without warning.

Family and friends of an epileptic individual should be aware of what to do if a seizure occurs. During a seizure the clothing should be loosened around the neck, the head should be cushioned with a pillow, and any sharp or hard objects should be removed from the area. An object should never be inserted into the person's mouth during a seizure. After the seizure the head of the individual should be turned to the side to drain secretions from the mouth.

Headache

There are four major varieties of headache, each with its own type and severity of pain, temporal relationship, site and pattern of radiation, nature of aggravating or relieving factors, and associated symptoms.

Vascular headaches include migraine and its variants as well as headaches due to abnormal stretching of the arterial walls in the cranium as a result of vessel-wall disease. Migraine headaches are extremely painful recurring headaches that are sometimes accompanied by nausea and vomiting; most migraine sufferers have a family history of the disorder. The pain is typically severe, throbbing or pounding, felt on one or both sides of the head, and aggravated by activity, noise, or bright lights. Visual phenomena such as zigzag lights may precede the onset of pain. Visual loss, weakness of a limb or of one side of the body, speech disturbances, or confusion may also accompany and outlast the headache. In a variant called cluster headache, severe pain is felt in and around one eye that lasts approximately an hour and frequently wakes the patient in the early morning. Such attacks occur about 10 to 20 times a month, with months of no symptoms until the next cluster begins. In the elderly, brief episodes of brain dysfunction resembling small strokes may occur, apparently caused by arterial spasm during migraines.

Foods such as cheese, chocolate, alcohol, and those containing nitrite or monosodium glutamate are the most frequent causes of migraines. Other causes include excessive or deficient sleep, stress, oral contraceptives, and menstruation. Medications such as ergotamine, antiemetic agents, and pain-relieving agents or even oxygen may be necessary to treat acute attacks. Preventive medications include tricyclics, adrenergic blocking agents, and lithium.

Tension headaches are continuous and generalized pains felt from front to back or all around the head; they are generally less severe than migraines. Although stress is the most common cause, arthritis of the neck may also cause such a headache. Tension headaches are often experienced by migraine sufferers, and a combined form of headache exists. Alleviation of stress may reduce symptoms.

Traction headaches are caused by the distortion of intracranial pain-sensitive structures. This type of headache may be caused by intracranial bleeding or tumours, brain infections, obstruction of the flow of cerebrospinal fluid, and low pressure of the fluid due to leakage after lumbar puncture. Such headaches are moderately but increasingly severe, involve the whole head, may be accompanied by a throbbing sensation, are worse with movement and coughing and on awaking, and improve as the day progresses but recur for hours.

Pain may also be referred to the head (i.e. felt in the head even though the site of disease is elsewhere) by eye disorders such as glaucoma, infections or tumours of the nasal sinuses, dental infections, and arthritis of the neck.

Infections

Encephalitis

Encephalitis, an infection of the brain, may be caused by a number of microorganisms including viruses, bacteria, and fungi. In the Western world, viral encephalitis is the most common type of the disorder; it is typically caused by the herpes simplex virus. Other causes of viral encephalitis are measles, mumps, polio, rabies, and influenza.

Typical symptoms of encephalitis include the onset, over hours or days, of headache, fever, stiffness of the neck, drowsiness, and malaise. CT scanning of the head, EEG, and lumbar puncture (spinal tap) may be necessary in order to diagnose the disorder. Herpes simplex encephalitis is treated with medications such as acyclovir or vidarabine.

Other Viral Diseases

Subacute sclerosing panencephalitis is characterized by the slowly increasing loss of mental abilities, brief, shock-like jerking of the body, weakness, and spasticity. The disorder primarily affects children and young people. The measles virus is the responsible agent, probably causing an abnormal response of the immune system.

Progressive multifocal leukoencephalopathy is another disease of the brain occurring in individuals whose immune system is suppressed by drugs or disease. Progressive loss of myelin occurs in the white matter of the brain, cerebellum, and spinal cord. The responsible agent is a polyoma virus.

Cerebral abscess is caused by bacterial infection that may be carried via the bloodstream in cases of generalized or distant infection or may result from infection following a skull fracture. Infection produces a pus-filled abscess, usually localized close to the site of the original infection. Symptoms resemble those of viral encephalitis. EEG and CT scan may be performed to localize and determine the source of the infection.

Cortical thrombophlebitis may result from an infection that travels along the course of the intracranial veins, which themselves are infected and thrombose, or clot. This disease can occur in the venous channels of the dura mater, resulting in infarcts of the surrounding brain and damage to contiguous cranial nerves. Cortical thrombophlebitis occurs primarily in children suffering from malnutrition and dehydration, but it also may result from some blood diseases and may occur in pregnancy. The symptoms resemble those of a cerebral abscess, but there is a greater likelihood of seizures.

Antibiotics are usually administered to treat cerebral abscesses and are chosen according to the infectious organism. Anticoagulants may be used to treat cortical thrombophlebitis. Surgery may be necessary to remove scar tissue from the abscess.

The infectious disease kuru was once prevalent in people of the Fore tribe of New Guinea. Transmission of the disease was traced to ritual cannibalism. Symptoms included abnormal involuntary movements, dementia, and disturbance of motor functions. The disease was invariably fatal. Thanks to the work of Stanley Prusiner in the 1980s, it was discovered that a transmissible agent he called a prion, a deviant form of a harmless protein normally found in the brain, causes the disease.

Prusiner's research also indicated that the abnormal protein causes the fatal degenerative disorder of the brain called Creutzfeldt–Jakob disease (CJD), scrapie, and other prion diseases by a catalytic process in which it, on contact with the normal protein, causes the latter to change its structure and become abnormal. In a chain reaction ever more of the abnormal protein is produced, and after months or years it finally accumulates to levels that cause obvious brain damage. In 1996, when a new variant of CJD emerged in the UK, Prusiner's research was the focus of national attention.

Prusiner's work could help scientists understand Alzheimer's disease and other more common brain disorders. For example, some researchers believed that Alzheimer's disease is caused by a structural change in certain non-prion proteins, which leads to the accumulation of abnormal deposits in the brain.

Prions

Prions are an abnormal form of a normally harmless protein found in the brain that is responsible for a variety of fatal neurodegenerative diseases of both animals and humans called transmissible spongiform encephalopathies.

In the early 1980s the American neurologist Stanley B. Prusiner and colleagues identified the "proteinaceous infectious particle", a name that was shortened to "prion" (pronounced "pree-on"). Prions can enter the brain through infection, or they can arise from mutations in the gene that encodes the protein. Once present in the brain prions multiply by inducing benign proteins to refold into the abnormal shape.

This mechanism is not fully understood, but another protein normally found in the body may also be involved.

The normal protein structure is thought to consist of a number of flexible coils called alpha helices. In the prion protein some of these helices are stretched into flat structures called beta strands. The normal protein conformation can be degraded rather easily by cellular enzymes called proteases, but the prion protein shape is more resistant to this enzymatic activity. Thus, as prion proteins multiply they are not broken down by proteases and instead accumulate within nerve cells, destroying them. Progressive nerve cell destruction eventually causes brain tissue to become filled with holes in a sponge-like, or spongiform, pattern.

Diseases caused by prions that affect humans include CJD, Gerstmann–Sträussler–Scheinker disease, fatal familial insomnia, and kuru. Prion diseases affecting animals include scrapie, bovine spongiform encephalopathy (BSE, commonly called mad cow disease), and chronic wasting disease of mule deer and elk. For decades physicians thought that these diseases resulted from infection with slow-acting viruses, so called because of the lengthy incubation times required for the illnesses to develop. These diseases were, and sometimes still are, referred to as slow infections. The pathogenic agent of these diseases does have certain viral attributes, such as extremely small size and strain variation, but other properties are atypical of viruses. In particular, the agent is resistant to ultraviolet radiation, which normally inactivates viruses by destroying their nucleic acid.

Prions are unlike all other known disease-causing agents in that they appear to lack nucleic acid (i.e. DNA or RNA), which is the genetic material that all other organisms contain. Another unusual characteristic of prions is that they can cause hereditary, infectious, and sporadic forms of disease – for example, CJD manifests in all three ways, with sporadic cases being the most common. Prion proteins can act as infectious agents, spreading disease when transmitted to another organism, or they can arise from an inherited mutation. Prion diseases also show a sporadic pattern of incidence, meaning that they seem to appear in the population at random. The underlying molecular process that causes the prion protein to form in these cases is unknown. Other neurodegenerative disorders, such as Alzheimer's or Parkinson's disease, may arise from molecular mechanisms similar to those that cause the prion diseases.

Down's Syndrome

Down's syndrome (also called Down syndrome, trisomy 21, or [formerly] Mongolism) is a congenital disorder caused by an extra chromosome on the chromosome 21 pair, thus giving the person a total of 47 chromosomes rather than the normal 46. Those born with Down's syndrome are characterized by several of the following: broad, flat face; short neck; up-slanted eyes, sometimes with an inner epicanthal fold; low-set ears; small nose and enlarged tongue and lips; sloping underchin; poor muscle tone; intellectual disability; heart or kidney malformations or both; and abnormal dermal ridge patterns on fingers, palms, and soles. The intellectual disability seen in those with Down's syndrome is usually moderate, though in some it may be mild or severe. Congenital heart disease is found in about 40 per cent of people with Down's syndrome.

Most people with Down's syndrome have an extra (third) chromosome – a condition known as trisomy – associated with the chromosome 21 pair. Yet while trisomy is the norm, a small number (perhaps four per cent) have an abnormality called translocation, in which the extra chromosome in the 21 pair breaks off and attaches itself to another chromosome. The cause of the chromosomal abnormalities in Down's syndrome remains unknown.

Down's syndrome occurs in about one in every 800 live births. The incidence of the disorder increases markedly in the offspring of women over the age of 35. This is illustrated by the fact that the incidence of Down's syndrome in the offspring of young women is only about one in 1,000, while its incidence in those of women over age 40 is about one in 40. Down's syndrome can be diagnosed prenatally by the presence of the abnormal chromosome in samples of fetal cells taken from the amniotic fluid.

With modern medical care, most people with Down's syndrome – except those with major heart defects that cannot be corrected by surgery – live into adulthood. They do have a shorter life expectancy (55) than normal adults, however, because they develop the degenerative conditions of old age prematurely. Because those with Down's syndrome have an intellectual disability to varying degrees, some never become self-supporting. Most can, however, be taught to contribute usefully in the home or in a sheltered working or living environment after they are grown.

Stroke

Blood follows two pathways to the brain: the two internal carotid arteries, which divide intracranially into a number of branches (the largest being the anterior and middle cerebral arteries); and the vertebral arteries, which fuse at the base of the brain and then divide again to form the posterior cerebral arteries. There are many communications between vessels inside and outside the head, and, in some places, two arteries supply the same territory. Despite this generous arterial network, blood supply is actually precarious due to the huge requirements of the brain for blood and the fact that the last branches of arteries anastomose, or join together, very little if at all. To help mediate the different rates of cerebral blood flow caused by variations in heartbeat, respiration, blood pressure, and posture, a system of autoregulation exists whereby the cerebral blood vessels vary their size in response to such changes. Yet, in diseased states, blood supply to parts of the brain still often fails. The effects vary from minor, transient slowing of activity, due to slight reduction in local blood flow, to strokes, which are severe and cause permanent loss of function when blood supply is completely lost.

The features of transient ischemic attacks (TIAs) and of any stroke syndrome depend on which part of the brain is affected. Transient loss of vision in one eye, as if a curtain were being pulled over it, is one common sign of ischaemia of the retina. Similar brief episodes of numbness or weakness of a limb or difficulty in speech suggest an attack in the carotid artery, while a brief reduction in consciousness, vertigo, slurred speech, impaired vision in both eyes, or imbalance may signify ischaemia in the vertebrobasilar circulation.

Infarction of the brain occurs most often during sleep; the individual awakes to find that they have lost the function of part of their body. The most common site of stroke is in the middle cerebral artery. The usual symptoms are aphasia and other disorders of higher cerebral functioning, hemiparesis (weakness of the face and arm on the other side of the body), and loss of sensation in the same areas. Blockage of the internal carotid artery in the neck produces similar symptoms. Because there is extensive communication between the anterior cerebral arteries on both sides, the area of the brain supplied by the anterior cerebral artery on the occluded side may escape damage. Anterior cerebral artery infarcts produce weakness and sensory changes on the opposite side of the body, but the leg is more affected than the arm. Posterior cerebral infarcts cause loss of vision in the opposite half of the visual field and sometimes difficulty in comprehending speech (called receptive aphasia).

Small, deep infarcts in the white matter cause small cavities of damaged tissue, called lacunes. Infarcts in this area cause a number of syndromes in which the damage is restricted to, for example, pure weakness or pure sensory dysfunction on one side of the body. Intracerebral haemorrhage most commonly occurs in the putamen and leads to sudden headache, severe hemiparesis, loss of consciousness, and deviation of the eyes to one side. Bleeding into the pons causes paresis of all limbs and unconsciousness. Bleeding into the cerebellum produces typical signs of incoordination with headache and stiffness of the neck. Subarachnoid haemorrhage (SAH) may be caused by rupture of an aneurysm and may cause a sudden headache, a stiff neck, a seizure, or loss of consciousness and, depending on the damage done by the blood pumped out of the artery, brain damage.

Occlusive Strokes

Occlusive strokes, those in which a blood vessel supplying a part of the brain is blocked, are divided into four groups. (1) Transient ischemic attacks (TIAs) are the mildest occlusive strokes; symptoms last for minutes or hours. TIAs are usually caused by small emboli, such as fragments composed of blood cells or cholesterol, that are swept into the circulation of the brain from the arteries of the neck or elsewhere. (2) Reversible ischemic neurological deficits may have the same pathological basis, but symptoms last for up to three weeks. (3) Evolving strokes are characterized by a stuttering or progressive deterioration. Blockage of the larger arteries in the brain or neck is usually responsible. (4) Completed strokes cause reduction in blood supply (ischemia) and death of brain tissue as a result of severe starvation of blood (infarction).

The most common pathological change causing stroke is atherosclerosis of the large- and medium-sized arteries in the neck and brain, which causes ulcers in the wall of the artery, narrowing of the lumen, or blockage of the vessel. Small-vessel disease is characterized by the thickening and degeneration of the walls of the arterioles and is usually due to high blood pressure. Other vessel-wall diseases include inflammatory changes, as in collagen-vascular diseases or after deep X-ray treatment; spasm, as in severe migraine headaches; splitting of the wall of the artery; and direct injury. Larger emboli causing more than just a TIA may arise in the larger arteries and from the valves or lining membrane of the heart.

Other causes of stroke are a severe global decrease in cerebral blood flow, as may occur with a severe haemorrhage or heart attack; diversion of blood to other arteries, as a result of a narrowing or blockage of a vertebral artery in the neck; cerebral anoxia, a severe reduction in the amount of oxygen in the blood; and abnormal viscosity of the blood due to excess numbers of cells or to increased protein concentration in the blood.

Haemorrhagic Strokes

Haemorrhagic strokes, in which a blocked vessel bleeds, are usually due to small-vessel disease in individuals with high blood pressure. Such intracerebral haemorrhage occurs most often in the deep white matter of the brain, brainstem, and cerebellum. The incidence of such strokes has been reduced in part by the introduction of more effective treatments for hypertension. Bleeding also occurs when a defect in the wall of an artery, called an aneurysm, ruptures. When this occurs at the sites of branching of the larger arteries inside the head, blood spills into the subarachnoid space, causing subarachnoid haemorrhage. Bleeding may also occur with malformations of arteries and veins, into an infarct, and with blood diseases that impair coagulation.

The diagnosis of cerebrovascular disease is determined by clinical evaluation, but it has been greatly improved by MRI and CT scanning, which not only show the site of the stroke but also allow determination of bleeding. Additional tests may include angiography, for localizing an aneurysm or detecting the location and nature of disease in the arteries of the neck, and a battery of tests that determine whether any underlying disease (e.g. diabetes, blood cell or clotting diseases, infections) might have led to the stroke.

Treatment of a stroke is initially directed toward support of the cardiac and metabolic systems. High blood pressure is controlled, raised intracranial pressure lowered, and, in some cases, anticoagulant medications prescribed. Rehabilitative efforts start as soon as the patient's condition is stable and may include cessation of smoking, diabetes treatment, clipping or isolation of a cerebral aneurysm, and, perhaps, surgical removal of a blood clot or the roughened lining of the major arteries in the neck. Medications such as aspirin, which inhibit the tendency of platelets to cause clotting in the bloodstream, have proved effective in the prevention of stroke and fatal heart attacks after TIAs in men.

Tumours

The direct effects of brain tumours are replacement or compression of tissue, which impairs the function of the compressed part and often causes epileptic seizures. Indirect effects of tumours on the brain include raised intracranial pressure, which is particularly common in the case of tumours arising within or compressing the ventricular system.

Benign Tumours

Benign intracranial tumours do not spread within the brain or metastasize to distant sites. The most common benign brain tumours are neurofibromas (tumours of the myelin-forming Schwann cells), tumours of the skull, and meningiomas. Pituitary adenomas arise within the pituitary fossa. By compressing the underside of the optic chiasm, these tumours cause visual deficits, and they raise the intracranial pressure through compression of the hypothalamus and third ventricle. In addition, many pituitary adenomas secrete hormones, which may stimulate such reactions as abnormal growth and lactation. Surgical removal through the nose or through craniotomy (opening up the cranial cavity), is often required; small tumours of the pituitary may be suppressed by the medication bromocriptine.

Craniopharyngiomas arise from a vestigial remnant of tissue in the roof of the mouth and extend upward to cause effects similar to those of pituitary tumours. Because they often contain calcium, craniopharyngiomas may be seen with X-rays. Craniotomy is necessary for removal.

Colloid cysts may be found within the ventricles. Although they seldom cause symptoms, on account of their small size, they occasionally may block the flow of cerebrospinal fluid, causing an acute rise in intracranial pressure that results in headache and loss of consciousness.

Other benign intracranial masses include parasitic cysts, granulomas, tuberculomas, and dermoid cysts. These masses may arise in any part of the brain or beside it, in which case there may also be involvement of the skull.

Malignant Tumours

Malignant tumours of the brain are more common than benign ones; the most frequent malignant brain tumours are gliomas, which arise from the neuroglial cells. Astrocytomas may arise anywhere in the brain and destroy the function of the tissues that they invade and replace; they also cause seizures and eventually signs of raised intracranial pressure. Treatment includes surgical removal and radiation therapy.

Medulloblastomas and ependymomas are other varieties of glioma arising, respectively, in the brainstem and in the walls of the ventricles. Ependymomas also grow in the posterior fossa in young people, but more slowly. These tumours tend to cause an increase in intracranial pressure and signs of brainstem and cerebellar dysfunction. Surgical removal is seldom complete, but radiation therapy may be able to shrink the tumour and slow its growth.

Oligodendrogliomas may occur anywhere in the cerebrum and are particularly liable to cause epileptic seizures. They often contain calcium, which may allow their detection with X-rays. Teratomas and pinealomas are more common in young men; arising from the pineal gland and damaging the hypothalamus and upper brainstem, they disturb the control of eye movement, obstruct the flow of cerebrospinal fluid, and cause cerebellar dysfunction. Their effects may resemble those of a benign or of a malignant tumour. Lymphomas such as

Hodgkin's disease rarely invade the brain, although they may compress it. Reticulum cell sarcomas (microgliomas) do invade the brain diffusely and may cause headache, seizures, personality changes, and various focal signs.

A history of headaches, seizures, and focal neurological problems may indicate a brain tumour; diagnosis is confirmed by MRI, CT scanning, or radioisotope scanning. Surgery is usually necessary in order to confirm the nature of the tumour, to relieve increased pressure, and to remove all or part of the mass. This also may make radiation therapy more effective. Swelling surrounding tumours can often be reduced by steroid medications.

Malignant tumours arising elsewhere in the body may metastasize to the brain; common sites of origin of such secondary tumours are the lung, breast, thyroid, and kidney. These are usually multiple tumours, so surgical removal is usually unsuccessful; however, radiation therapy may shrink them and slow their growth.

Brain Cancer

Brain cancer, the result of malignant tumour, is caused by the uncontrolled growth of cells in the brain. The term refers to any of a variety of tumours affecting different brain cell types. Depending on the location and cell type, brain cancers may progress rapidly or slowly over a period of many years. Brain cancers are often difficult to treat, and complete cure is often unattainable.

The causes of different brain cancers remain largely unknown. Unlike many other cancers, brain tumours seem to occur at random in the population and are not usually associated with known risk factors. However, exposure to ionizing radiation, such as during head X-rays, does increase a person's risk of developing certain brain cancers, as does a suppressed immune system or family history of cancer. Symptoms of brain cancer vary widely depending on the location of the tumour. As the tumour grows, it might put pressure on nearby regions of the brain and thereby affect the functions controlled by those regions. Difficulty or changes in speech, hearing, vision, or motor functions can all indicate the presence of a brain tumour. Many brain tumours are initially discovered following chronic headaches, and in some cases seizures are associated with cancers of the brain. Symptoms may also include vomiting, nausea, or numbness in any part of the body.

If a brain tumour is suspected, a neurological exam is conducted to test general brain function. Further diagnosis usually utilizes imaging procedures such as X-rays, CT scans, and MRI. The location and stage of a tumour can also be determined with PET scans. The blood supply feeding a tumour can be assessed by using an X-ray procedure called angiography. A definitive diagnosis usually requires removal of brain tissue for analysis; often this is done during tumour-removal surgery. In other cases, a needle biopsy guided by the images generated by CT scans or MRI may be used to access the tumour.

Brain cancers are usually not diagnosed until symptoms have appeared, and survival rates vary widely, depending on type and location. Some are completely curable. Slow-growing cancers may progress for decades, whereas other types may be fatal within six to eight years. Average survival from some faster-growing tumours, however, averages no more than one year.

Surgery is the most frequent approach to treating brain tumours. Such surgery may be curative for some cancers, but for others it may only relieve symptoms and prolong survival. In many cases, complete removal of the tumour is not possible.

Radiation therapy may be used to cure some brain cancers,

but others do not respond to radiotherapy. Radiation generally works best with fast-growing types. Because radiation therapy typically involves X-rays, which pose a risk to healthy brain tissue, it is important to minimize exposure to the normal cells surrounding the tumour. This is accomplished by employing special procedures that focus the radiation. For instance, a device called a gamma knife, which emits a highly controllable beam of radiation, may be used. Even when radiation is localized, however, radiotherapy can cause side-effects such as vomiting, diarrhoea, or skin irritation. Radiation to the brain may cause scar tissue to form and potentially cause future problems. Memory loss may also occur.

Chemotherapy is used for some brain tumours, but, owing to the brain's protective barrier, many chemotherapeutic agents cannot enter the brain from the bloodstream. Chemotherapy works best on fast-growing tumours, but it is generally not curative and causes side-effects similar to radiation therapy. Both radiation therapy and chemotherapy are often used when someone's general health or the location of the tumour prevents surgery.

In rare cases where a family history or a personal history of frequent head X-rays suggests an increased risk of brain cancer, regular screening by a neurologist may allow developing cancers to be detected earlier. Otherwise, no means of preventing brain tumours are known.

Meningitis

Meningitis is an inflammation of the meningeal coverings of the nervous system, with possible involvement of the brain. Viruses such as mumps, Coxsackie, and ECHO viruses, tuberculosis, fungi, spirochetes, bacteria, protozoa, and some chemical agents may cause the disease. Organisms most often reach the meninges via the blood, but direct spread may occur with skull fractures, middle-ear or nasal-sinus infections, or congenital defects of the meninges. Symptoms depend upon the infectious organism and the resistance and age of the patient, but they usually include lethargy and drowsiness, fever, headache, stiffness of the neck, vomiting, and (in smaller children) seizures. Patients with non-bacterial, or aseptic, meningitis also have fever, headache, and other meningeal signs, but they are not so obviously ill. Residual consequences of meningitis include cranial nerve palsies (especially loss of hearing), hydrocephalus, and brain damage.

Because treatment with the correct antibiotic is essential to treat meningitis, the most important diagnostic aid is lumbar puncture, the examination of the cerebrospinal fluid. The amount of protein, the number and type of cells, and the glucose level of the fluid confirm the type of meningitis.

11

MENTAL DISORDERS

Mental disorder is any illness with significant psychological or behavioural manifestations that is associated with either a painful or distressing symptom or an impairment in one or more important areas of functioning.

Mental disorders, in particular their consequences and their treatment, are of more concern and receive more attention now than in the past. Mental disorders have become a more prominent subject of attention for several reasons. They have always been common, but, with the eradication or successful treatment of many of the serious physical illnesses that formerly afflicted humans, mental illness has become a more noticeable cause of suffering and accounts for a higher proportion of those disabled by disease. Moreover, the public has come to expect the medical and mental health professions to help it obtain an improved quality of life in its mental as well as physical functioning. And indeed, there has been a proliferation of both pharmacological and psychotherapeutic treatments. The transfer of many psychiatric patients, some still showing conspicuous symptoms, from mental hospitals into the community has also increased the public's awareness of the importance and prevalence of mental illness.

There is no simple definition of mental disorder that is universally satisfactory. This is partly because mental states or behaviour that are viewed as abnormal in one culture may be regarded as normal or acceptable in another, and in any case it is difficult to draw a line clearly demarcating healthy from abnormal mental functioning.

A narrow definition of mental illness would insist upon the presence of organic disease of the brain, either structural or biochemical. An overly broad definition would define mental illness as simply being the lack or absence of mental health – that is to say, a condition of mental well-being, balance, and resilience in which the individual can successfully work and function and in which the individual can both withstand and learn to cope with the conflicts and stresses encountered in life. A more generally useful definition ascribes mental disorder to psychological, social, biochemical, or genetic dysfunctions or disturbances in the individual.

A mental illness can have an effect on every aspect of a person's life, including thinking, feeling, mood, and outlook and such areas of external activity as family and marital life, sexual activity, work, recreation, and management of material affairs. Most mental disorders negatively affect how individuals feel about themselves and impair their capacity for participating in mutually rewarding relationships.

Psychopathology is the systematic study of the significant causes, processes, and symptomatic manifestations of mental disorders. The meticulous study, observation, and inquiry that characterize the discipline of psychopathology are, in turn, the basis for the practice of psychiatry (i.e. the science and practice of diagnosing and treating mental disorders as well as dealing with their prevention). Psychiatry, psychology, and related disciplines such as clinical psychology and counselling embrace a wide spectrum of techniques and approaches for treating mental illnesses. These include the use of psychoactive drugs to correct biochemical imbalances in the brain or otherwise to relieve depression, anxiety, and other painful emotional states.

Another important group of treatments is the psychotherapies which seek to treat mental disorders by psychological means and which involve verbal communication between the patient and a trained person in the context of a therapeutic interpersonal relationship between them. Different modes of psychotherapy focus variously on emotional experience, cognitive processing, and overt behaviour.

Psychiatric classification attempts to bring order to the enormous diversity of mental symptoms, syndromes, and illnesses that are encountered in clinical practice. Epidemiology is the measurement of the prevalence, or frequency of occurrence, of these psychiatric disorders in different human populations.

Psychoses

Psychoses are major mental illnesses characterized by severe symptoms such as delusions, hallucinations, disturbances of the thinking process, and defects of judgement and insight. People with psychoses exhibit a disturbance or disorganization of thought, emotion, and behaviour so profound that they are often unable to function in everyday life and may be incapacitated or disabled. Such individuals are often unable to realize that their subjective perceptions and feelings do not correlate with objective reality, a phenomenon evinced by people with psychoses who do not know or will not believe that they are ill despite the distress they feel and their obvious confusion concerning the outside world. Traditionally, the psychoses have been broadly divided into organic and functional psychoses. Organic psychoses were believed to result from a physical defect of or damage to the brain. Functional psychoses were believed to have no physical brain disease evident upon clinical examination. Much recent research suggests that this distinction between organic and functional is probably inaccurate. Most psychoses are now believed to result from some structural or biochemical change in the brain.

Schizophrenia

The term schizophrenia was introduced by Swiss psychiatrist Eugen Bleuler in 1911 to describe what he considered to be a group of severe mental illnesses with related characteristics; it eventually replaced the earlier term dementia praecox, which the German psychiatrist Emil Kraepelin had first used in 1899 to distinguish the disease from what is now called bipolar disorder. Individuals with schizophrenia exhibit a wide variety of symptoms; thus, although different experts may agree that a particular individual suffers from the condition, they might disagree about which symptoms are essential in clinically defining schizophrenia.

The annual prevalence of schizophrenia – the number of cases, both old and new, on record in any single year – is between two and four per 1,000 people. The lifetime risk of developing the illness is between seven and nine per 1,000. Schizophrenia is the single largest cause of admissions to mental hospitals, and it accounts for an even larger proportion of the permanent populations of such institutions. It is a severe and frequently chronic illness that typically first manifests itself during the teen years or early adulthood. More severe levels of impairment and personality disorganization occur in schizophrenia than in almost any other mental disorder.

The principal clinical signs of schizophrenia may include delusions, hallucinations, a loosening or incoherence of a person's thought processes and train of associations, deficiencies in feeling appropriate or normal emotions, and a withdrawal from reality. A delusion is a false or irrational belief that is firmly held despite obvious or objective evidence to the contrary. The delusions of individuals with schizophrenia may be persecutory, grandiose, religious, sexual, or hypochondriacal in nature, or they may be concerned with other topics. Delusions of reference, in which the person attributes a special, irrational, and usually negative significance to other people, objects, or events, are common in the disease. Especially characteristic of schizophrenia are delusions in which the individual believes their thinking processes, body parts, or actions or impulses are controlled or dictated by some external force.

Hallucinations are false sensory perceptions that are experienced without an external stimulus but that nevertheless seem real to the person who is experiencing them. Auditory hallucinations, experienced as "voices" and characteristically heard commenting negatively about the affected individual in the third person, are prominent in schizophrenia. Hallucinations of touch, taste, smell, and bodily sensation may also occur. Disorders of thinking vary in nature but are quite common in schizophrenia. Thought disorders may consist of a loosening of associations, so that the speaker jumps from one idea or topic to another, unrelated one in an illogical, inappropriate, or disorganized way. At its most serious, this incoherence of thought extends into pronunciation itself, and the speaker's words become garbled or unrecognizable. Speech may also be overly concrete and inexpressive; it may be repetitive, or, though voluble, it may convey little or no real information. Usually individuals with schizophrenia have little or no insight into their own condition and realize neither that they are suffering from a mental illness nor that their thinking is disordered.

Among the so-called negative symptoms of schizophrenia are a blunting or flattening of the person's ability to experience (or at least to express) emotion, indicated by speaking in a monotone and by a peculiar lack of facial expressions. The person's sense of self (i.e. of who they are) may be disturbed. Someone with schizophrenia may be apathetic and may lack the drive and ability to pursue a course of action to its logical conclusion, may withdraw from society, become detached from others, or become preoccupied with bizarre or nonsensical fantasies. Such symptoms are more typical of chronic rather than of acute schizophrenia.

Experts have recognized different types of schizophrenia as well as intermediate stages between the disease and other conditions. Five major types of schizophrenia are recognized by the Diagnostic and Statistical Manual of Mental Disorders: the disorganized type, the catatonic type, the paranoid type, the undifferentiated type, and the residual type. Disorganized schizophrenia is characterized by inappropriate emotional responses, delusions or hallucinations, uncontrolled or inappropriate laughter, and by incoherent thought and speech. Catatonic schizophrenia is marked by striking motor behaviour, such as remaining motionless in a rigid posture for hours or even days, and by stupor, mutism, or agitation. Paranoid schizophrenia is characterized by the presence of prominent delusions of a persecutory or grandiose nature; some patients can be argumentative or violent. The undifferentiated type combines symptoms from the above three categories, while the residual type is marked by the absence of these distinct features; moreover, the residual type, in which the major symptoms have abated, is a less severe diagnosis.

Autism

Autism is a developmental disorder that affects physical, social, and language skills. The syndrome usually appears before three years of age, although the earliest signs are quite subtle.

Autistic children appear indifferent or averse to affection and physical contact, although attachment to parents or certain adults often develops later. Speech develops slowly and abnormally (it is often atonal and arrhythmic) or not at all. It may be characterized by meaningless, non-contextual echolalia (constant repetition of what is said by others) or the replacement of speech by strange mechanical sounds. There may be abnormal reaction to sound, no reaction to pain, or no recognition of genuine danger, yet autistic children are extremely sensitive. Usually the syndrome is accompanied by an obsessive desire to prevent environmental change. Frequently there are also rhythmic body movements, such as rocking or hand-clapping. About 25 per cent of autistic children develop seizures by late adolescence.

Estimates of the prevalence of autism range from 10 to 20 per 10,000 children; some 15 to 20 per cent are able to become socially and vocationally independent. The disorder is about four times more common in males.

Autism is still incompletely understood. Abnormalities of the brain (particularly in the cerebellum, brainstem, and limbic system) are likely to have occurred during early development. Genetic or environmental influences, a deficiency of large neurons called Purkinje cells in the cerebellum, or an excess of the neurotransmitter serotonin may also cause autism.

There is no cure for autism; behavioural or drug therapy may improve some symptoms. People with the condition have a normal life expectancy.

Neuroses

Neuroses, or psychoneuroses, are less serious disorders in which people may experience negative feelings such as anxiety or depression. Their functioning may be significantly impaired, but personality remains relatively intact, the capacity to recognize and objectively evaluate reality is maintained, and they are basically able to function in everyday life. In contrast to people with psychoses, neurotic patients know or can be made to realize that they are ill, and they usually want to get well and return to a normal state. Their chances for recovery are better than those of people with psychoses. The symptoms of neurosis may sometimes resemble the coping mechanisms used in everyday life by most people, but in neurotics these defensive reactions are inappropriately severe or prolonged in response to an external stress. Anxiety disorders, phobic disorder (exhibited as unrealistic fear or dread), conversion disorder (formerly known as hysteria), obsessive–compulsive disorder (OCD), and depressive disorders have been traditionally classified as neuroses.

Neuroses are therefore characterized by anxiety, depression, or other feelings of unhappiness or distress that are out of proportion to the circumstances of a person's life. They may impair a person's functioning in virtually any area of their life, relationships, or external affairs, but they are not severe enough to incapacitate the person. Neurotic patients generally do not suffer from the loss of the sense of reality seen in people with psychoses.

Psychiatrists first used the term neurosis in the mid-nineteenth century to categorize symptoms thought to be neurological in origin; the prefix "psycho-" was added some decades later when it became clear that mental and emotional factors were important in the aetiology of these disorders. The terms are now used interchangeably, although the shorter word is more common. Both terms, however, lack the precision required for psychological diagnosis and are no longer used for that purpose.

An influential view held by the psychoanalytic tradition is that neuroses arise from intrapsychic conflict (conflict between different drives, impulses, and motives held within various components of the mind). Central to psychoanalytic theory, which was founded by Austrian neurologist Sigmund Freud, is the postulated existence of an unconscious part of the mind which, among other functions, acts as a repository for repressed thoughts, feelings, and memories that are disturbing or otherwise unacceptable to the conscious mind. These repressed mental contents are typically sexual or aggressive urges or painful memories of an emotional loss or an unsatisfied longing dating from childhood. Anxiety arises when these unacceptable and repressed drives threaten to enter consciousness; prompted by anxiety, the conscious part of the mind (the ego) tries to deflect the emergence into consciousness of the repressed mental contents through the use of defence mechanisms such as repression, denial, or reaction formation. Neurotic symptoms often begin when a previously impermeable defence mechanism breaks down and a forbidden drive or impulse threatens to enter consciousness.

While the psychoanalytic theory of neurosis has continued to be influential, another prominent view, associated with behavioural psychology, represents neurosis as a learned, inappropriate response to stress that can be unlearned. A third view, stemming from cognitive theory, emphasizes the way in which maladaptive thinking – such as the fear of possible punishment – promotes an inaccurate perception of the self and surrounding events.

Obsessive–Compulsive Disorders

Obsessive–compulsive disorder (OCD, also called obsessive–compulsive neurosis) is a type of mental disorder in which an individual experiences obsessions or compulsions or both. Either the obsessive thought or the compulsive act may occur singly, or both may appear in sequence.

Obsessions are recurring or persistent thoughts, images, or impulses that, rather than being voluntarily produced, seem to invade a person's consciousness despite their attempts to ignore, suppress, or control them. Obsessional thoughts are frequently morbid, shameful, repugnant, or merely tedious; they are usually experienced as being meaningless and are accompanied by anxiety to a varying degree. Common obsessions include thoughts about committing violent acts, worries about contamination (as by shaking hands with someone), and doubt (as in wondering whether one had turned off the stove before leaving the house).

Obsessions are accompanied by compulsions in approximately 80 per cent of cases. Compulsions are urges or impulses to commit repetitive acts that are apparently meaningless, stereotyped, or ritualistic. The compulsive person may be driven to perform the act not as an end in itself but as a means to produce or prevent some other situation, although they are usually aware that the two bear no logical causal relation to each other. Most compulsive acts are rather simple – such as persistent hand washing, counting, checking (e.g. the turned-off stove), touching, or the repetition of stereotyped words or phrases. Occasionally, however, elaborately for-malized and time-consuming ceremonials are necessary. The compulsive person usually knows the act to be performed is meaningless, but their failure or refusal to execute it brings on a mounting anxiety that is relieved once the act is performed. Should the sufferer be forcibly or externally prevented from performing the compulsive act, they may experience an over-whelming anxiety.

OCDs affect from two to three per cent of the general population, occur equally in males and females, and can first appear at any age. The tricyclic antidepressant (TCA) drug clomipramine (Anafranil) and the selective serotonin reuptake inhibitor (SSRI) fluoxetine (Prozac) have been found to mark-edly reduce the symptoms in about 60 per cent of cases and have thus become the treatment of choice. Both drugs affect the brain's metabolism of the neurotransmitter serotonin, and this had led researchers to suspect that OCDs arise primarily from defects in the brain's neurochemical functioning rather than from purely psychological causes. A drug traditionally used for tuberculosis, D-cycloserine (Seromycin), has also been shown, when used in combination with behavioural therapy, to increase the rate of fear extinction in patients with OCD. The highest rates of the condition occur in high-stress groups, such as those who are young, divorced, or unemployed.

Attention-Deficit Disorders

Characterized by inattention and distractibility, restlessness, inability to sit still, and difficulty concentrating on one thing for any period of time, ADHD most commonly occurs in children, although an increasing number of adults are being diagnosed with the disorder. ADHD is three times more common in males than in females and occurs in approximately three to six per cent of all children. Although behaviours characteristic of the syndrome are evident in all cultures, they have garnered the most attention in the USA, where ADHD is the most commonly diagnosed childhood psychiatric disorder.

It was not until the mid-1950s that American physicians began to classify as "mentally deficient" individuals who had difficulty paying attention on demand. Various terms were coined to describe this behaviour, among them minimal brain damage and hyperkinesis. In 1980 the American Psychiatric Association (APA) replaced these terms with attention deficit disorder (ADD). Then in 1987 the APA linked ADD with hyperactivity, a condition that sometimes accompanies attention disorders but may exist independently. The new syndrome was named attention-deficit/hyperactivity disorder, or ADHD.

ADHD does not have easily recognizable symptoms or definitive diagnostic tests. Children and adults are diagnosed with ADHD if they persistently show a combination of traits including, among others, forgetfulness, distractibility, fidgeting, restlessness, impatience, difficulty sustaining attention in work, play, or conversation, or difficulty following instructions and completing tasks. According to criteria issued by the APA, at least six of these traits must be present "to a degree that is maladaptive", and these behaviours must cause "impairment" in two or more settings, e.g. at school, work, or at home. Inattention predominates in some cases, hyperactivity in others, and in a combined type of ADHD the two are present together. Studies have shown that more than a quarter of children with ADHD are held back a grade in school, and a third fail to graduate from high school. The learning difficulties associated with ADHD, however, should not be confused with a deficient intelligence.

The cause of ADHD is not known and may be a combination of both inherited and environmental factors. Many theories regarding causation have been abandoned for lack of evidence. Past suspects have included bad parenting; brain damage due to head trauma, infection, or exposure to alcohol or lead; food allergy; and too much sugar. ADHD is thought to be at least partly hereditary. About 40 per cent of children with the condition have a parent who has ADHD, and 35 per cent have a sibling who is affected.

Using imaging technologies such as PET and fMRI, neuro-biologists have found subtle differences in the structure and function of the brains of people with and without ADHD. One study, which compared the brains of boys with and without ADHD, found that the corpus callosum, the band of nerve fibres that connects the two hemispheres of the brain, contained slightly less tissue in those with ADHD. A similar study discovered small size discrepancies in the brain structures known as the caudate nuclei. In boys without ADHD, the right caudate nucleus was normally about three per cent larger than the left caudate nucleus; this asymmetry was absent in boys with ADHD.

Other studies have detected not just anatomic but functional differences between the brains of those with and without ADHD. One research team observed decreased blood flow through the right caudate nucleus in adults with ADHD. Another study showed that an area of the prefrontal cortex known as the left anterior frontal lobe metabolizes less glucose in adults with ADHD, an indication that this area may be less active than in those without ADHD. Still other research showed higher levels of the neurotransmitter norepinephrine throughout the brains of people with ADHD and lower levels of another substance that inhibits the release of norepineph-rine. Metabolites, or broken-down products, of another neu-rotransmitter, dopamine, have also been found in elevated concentrations in the cerebrospinal fluid of boys with ADHD.

These anatomical and physiological variations may all affect a sort of "braking system" in the brain. The brain is constantly coursing with many overlapping thoughts, emotions, impulses, and sensory stimuli. Attention can be defined as the ability to focus on one stimulus or task while resisting focus on the extraneous impulses; people with ADHD may have reduced ability to resist focus on these extraneous stimuli. The cortical-striatal-thalamic-cortical circuit, a chain of neurons in the brain that connects the prefrontal cortex, the basal ganglia, and the thalamus in one continuous loop, is thought to be one of the main structures responsible for impulse inhibition. The size and activity differences found in the prefrontal cortex and basal ganglia of people with ADHD may be evidence of a delay in the normal growth and development of this inhibitory circuit. If this supposition is true, it would help explain why the symptoms of ADHD sometimes subside with age. The cortical-striatal-thalamic-cortical circuit in the brains of people with ADHD may not fully mature – providing more normal levels of impulse inhibition – until the third decade of life, and it may never do so in some people. This developmental lag may also explain why stimulant medications work to enhance attention. In one study, treatment with Ritalin restored average levels of blood flow through the caudate nucleus; and in other trials dopamine levels, which normally decrease with age but remain high in people with ADHD, fell after treatment with Ritalin. The hypothesis would coincide, finally, with observations that the social development of children with ADHD progresses at the same rate as that of their peers but with a lag of two to three years.

The most common medication used to treat ADHD is methylphenidate (Ritalin), a mild form of amphetamine. Amphetamines increase the amount and activity of the neurotransmitter norepinephrine (noradrenaline) in the brain. Although such drugs act as a stimulant in most people, they have the paradoxical effect of calming, focusing, or "slowing down" people with ADHD. Ritalin was developed in 1955, and the number of children with ADHD taking this and related medications has increased steadily ever since. Between 1990 and 1996 alone, the number of American children regularly taking Ritalin grew from 500,000 to 1,300,000, according to one study. Another study found that Ritalin prescriptions for adults rose from 217,000 in 1992 to 729,000 in 1997. The fact that many people diagnosed with ADHD experience fewer problems once they start taking stimulants such as Ritalin may confirm a neurological basis for the condition. Ritalin and similar medications help people with ADHD to concentrate better, which helps them get more work done and, in turn, reduces frustration and increases self-confidence.

ADHD has been a subject of great controversy and debate. A number of people who have been diagnosed with the syndrome – some of them psychologists and psychiatrists – have challenged the notion that personality traits such as inattentiveness, impulsivity, and distractibility deserve the label 'symptoms'. They contend that many people labelled as having ADHD are neither "deficient" nor "disordered" – they are simply different. ADHD, this vocal minority argues, is not a mental disorder at all but a different state of mind, and it is because of this difference that people with ADHD often do not function well in standard learning or work environments. It is society and its expectations that have to change, they claim, not those with short attention spans and high energy.

Indeed, the view of ADHD as a problem requiring medical intervention is highly culture-bound, being largely peculiar to the USA and Canada. This is not to say that the behaviours characteristic of ADHD are absent from children in other nations. The larger question is whether children in other countries are identified by their parents, teachers, and physicians as having a problem. In Great Britain and France only about one per cent of children are diagnosed with "hyperkinetic disorder", the closest equivalent to ADHD in the World Health Organization's International Classification of Diseases (the diagnostic system used by most medical professionals outside North America). And the British medical establishment hopes this number will remain comparatively low. The British Psychological Society suggested in a 1997 report that physicians and psychiatrists should not follow the American example of applying medical labels to such a wide variety of attention-related disorders: "The idea that children who don't attend or who don't sit still in school have a mental disorder is not entertained by most British clinicians."

Emerging scientific evidence about the causes and consequences of ADHD lends some plausibility to this viewpoint. As noted above, neurologists are finding that the anatomical and physiological differences underlying ADHD appear to be mere variations in the timing of brain development, not outright defects. Other researchers suggest that the behaviours characteristic of ADHD may once have conferred an evolutionary advantage, which would explain why their underlying genetic components have been conserved in the human gene pool.

Nevertheless, the majority of American medical professionals are certain that ADHD is a disorder and not just a normal variance. Indeed, some argue that the categorization of ADHD as a neurobiological disorder was an important step forward, because it clearly distinguished the ability to pay attention or control one's impulses from other mental capacities such as innate intelligence. Once ADHD was acknowledged as a disorder, impulsive or inattentive people could no longer be dismissed as "slow" or "stupid". Instead, the disorder could be managed with an appropriate treatment regimen – usually including medication but also incorporating certain organizing techniques – that would allow a person with ADHD to develop to the full extent of their intelligence.

Somatoform Disorders

Somatoform disorders, which include the so-called hysterical, or conversion, neuroses, manifest themselves in physical symptoms such as blindness, paralysis, or deafness that are not caused by organic disease. Hysteria was among the earliest syndromes to be understood and treated by psychoanalysts, who believe that such symptoms result from fixations or arrested stages in an individual's early psychosexual development.

Anxiety Disorders

A feeling of dread, fear, or apprehension, often with no clear justification, anxiety is distinguished from fear because the latter arises in response to a clear and actual danger, such as one affecting someone's physical safety. Anxiety, by contrast, arises in response to apparently innocuous situations or is the product of subjective, internal emotional conflicts the causes of which may not be apparent to the person themselves. Some anxiety inevitably arises in the course of daily life and is considered normal. But persistent, intense, chronic, or recurring anxiety not justified in response to real-life stresses is usually regarded as a sign of an emotional disorder. When such an anxiety is unreasonably evoked by a specific situation or object, it is known as a phobia. A diffuse or persistent anxiety associated with no particular cause or mental concern is called general, or free-floating, anxiety.

There are many causes (and psychiatric explanations) for anxiety. Austrian neurologist Sigmund Freud viewed anxiety as the symptomatic expression of the inner emotional conflict caused when someone suppresses (from conscious awareness) experiences, feelings, or impulses that are too threatening or disturbing to live with. Anxiety is also viewed as arising from threats to an individual's ego or self-esteem, as in the case of inadequate sexual or job performance. Behavioural psychologists view anxiety as a learned response to frightening events in real life; the anxiety produced becomes attached to the surrounding circumstances associated with that event, so that those circumstances come to trigger anxiety in the person independently of any frightening event. Personality and social psychologists have noted that the mere act of evaluating stimuli as threatening or dangerous can produce or maintain anxiety.

An anxiety disorder may develop where anxiety is insufficiently managed, characterized by a continuing or periodic state of anxiety or diffuse fear that is not restricted to definite situations or objects. The tension is frequently expressed in the form of insomnia, outbursts of irritability, agitation, palpitations of the heart, and fears of death or insanity. Fatigue is often experienced as a result of excessive effort expended in managing the distressing fear. Occasionally the anxiety is expressed in a more acute form and results in physiological symptoms such as nausea, diarrhoea, urinary frequency, suffocating sensations, dilated pupils, perspiration, or rapid breathing. Similar indications occur in several physiological disorders and in normal situations of stress or fear, but they may be considered neurotic when they occur in the absence of any organic defect or pathology and in situations that most people handle with ease.

Other anxiety disorders include panic disorder, agoraphobia, stress and post-traumatic stress disorder (PTSD), obsessive–compulsive disorder OCD, and generalized anxiety.

Post-Traumatic Stress Disorder

Post-traumatic stress disorder (PTSD, also called post-traumatic stress syndrome), is an emotional condition that sometimes follows a traumatic event, particularly an event that involves actual or threatened death or serious bodily injury to oneself or others and that creates intense feelings of fear, helplessness, or horror. The symptoms of PTSD include the re-experiencing of the trauma either through upsetting thoughts or memories or, in extreme cases, through a flashback in which the trauma is relived at full emotional intensity. People with PTSD often report a general feeling of emotional numbness, experience increased anxiety and vigilance, and avoid reminders of the trauma, such as specific situations, thoughts, and feelings. It is normal to experience such reactions to some extent following trauma, and they are not considered symptoms of PTSD unless they last for at least one month or have a delayed onset. People with PTSD can also suffer from other psychological problems, particularly depression, anxiety, and drug abuse.

The experience of traumatic stress is very common, and an estimated 10 per cent of women and five per cent of men experience PTSD at some point in their life. The risk for developing PTSD varies greatly with different kinds of trauma. Women are especially likely to develop PTSD following rape or other forms of sexual assault. Combat exposure has been found to be the most common cause of PTSD in men in the USA. A major disaster or traumatic event can cause PTSD on a large scale. For example, in the immediate aftermath of the September 11 attacks on the World Trade Center, 7.5 per cent of New Yorkers who lived in Manhattan below 110th Street – that is, in the general area of the World Trade Center – were found to be suffering from PTSD. The disorder is most likely to develop among people who suffer the greatest exposure to the trauma, who have the least social support, and who fail to allow themselves to experience their difficult feelings and find a new way of eventually understanding their experience.

Studies employing PET and fMRI have shown that people with symptoms of PTSD have altered activity in the brain, primarily in the regions of the medial prefrontal cortex, thalamus, and anterior cingulate gyrus. This altered activity may facilitate and reinforce the brain's ability to recall specific traumatic memories, thereby making it difficult for people with PTSD to break the pattern of negative memory recall.

About 12.5 per cent of people with PTSD have increased levels of a kinase (a type of regulatory enzyme) called CDK5 (cyclin-dependent kinase 5). Normally, CDK5 works with other proteins in nerve cells to regulate brain development, and its absence has been shown to facilitate the elimination of memories associated with fear. In people with PTSD, the elevated levels of CDK5 may interfere with and prevent fear extinction and delay the ability to control emotional states and reactions when a traumatic memory is recalled.

Some professionals believe that PTSD following a traumatic event can be reduced by early psychological interventions that encourage a sharing of emotional experiences concerning the event. However, scientific research has shown that these interventions offer little help and may even exacerbate the disorder. Once an individual has developed PTSD, the two most effective treatments are antidepressant medication and trauma re-exposure. Trauma re-exposure is a form of directive psychotherapy that involves encouraging the victim to recount the trauma and, through gradual re-exposure to the trauma in memory, change their emotional reactions in an effort to come to a new understanding of the experience.

Depression

Depression is defined as a mood or emotional state that is marked by feelings of low self-worth or guilt and a reduced ability to enjoy life. Someone who is depressed usually experiences several of the following symptoms: feelings of sadness, hopelessness, or pessimism; lowered self-esteem and heightened self-depreciation; a decrease or loss of ability to take pleasure in ordinary activities; reduced energy and vitality; slowness of thought or action; loss of appetite; and disturbed sleep or insomnia. Depression differs from simple grief or mourning, which are appropriate emotional responses to the loss of loved ones or objects. Where there are clear grounds for someone's unhappiness, depression is considered to be present if the depressed mood is disproportionately long or severe vis-à-vis the precipitating event. Someone who experiences alternating states of depression and mania (abnormal elevation of mood) or hypomania (distinct, though not necessarily abnormal, elevation of mood) is said to suffer from bipolar disorder.

Depression is probably the most common psychiatric complaint and has been described by physicians since before the time of Hippocrates, who called it melancholia. The course of the disorder varies from person to person; it may be mild or severe, acute or chronic. Untreated, depression may last an average of four months or longer. Depression is twice as prevalent in women as in men. The typical age of onset is in the 20s, but it may occur at any age.

Depression can have many causes. Unfavourable life events can increase a person's vulnerability to depression or trigger a depressive episode. Negative thoughts about oneself and the world are also important in producing and maintaining depressive symptoms. However, both psychosocial and biochemical mechanisms seem to be important causes; the chief biochemical cause appears to be the defective regulation of one or more naturally occurring neurotransmitters in the brain, particularly norepinephrine and serotonin. Reduced quantities or reduced activity of these chemicals in the brain is thought to cause the depressed mood in some sufferers.

There are three main treatments for depression. The two most important – and widespread by far – are psychotherapy and psychotropic medication, specifically antidepressants. Psychotherapy aims to alter the patient's maladaptive cognitive and behavioural responses to stressful life events while also giving emotional support to the patient. Antidepressant medications, by contrast, directly affect the chemistry of the brain and presumably achieve their therapeutic effects by correcting the chemical dysregulation that is causing the depression. Two types of medications, tricyclic antidepressants and the more recently developed selective serotonin reuptake inhibitors (SSRIs), though chemically different, both serve to prevent the presynaptic reuptake of serotonin (and in the case of tricyclic antidepressants, norepinephrine as well). This results in the build-up or accumulation of neurotransmitters in the brain and allows them to remain in contact with the nerve cell receptors longer, thus helping to elevate the patient's mood. By contrast, the antidepressants known as monoamine oxidase inhibitors (MAO) interfere with the activity of monoamine oxidase, an enzyme that is known to be involved in the breakdown of norepinephrine and serotonin. In cases of severe depression in which therapeutic results are needed quickly, electroconvulsive therapy (ECT) has sometimes proved helpful. In this procedure, a convulsion is produced by passing an electric current through the person's brain. In many cases of treatment, the best therapeutic results are obtained by using a combination of psychotherapy and antidepressant medication (see page 272).

Bipolar Disorder

Bipolar disorder was formerly called manic depression or manic-depressive illness. It is a mental disorder characterized by severe and recurrent depression or mania with abrupt or gradual onsets and recoveries. The states of mania and depression may alternate cyclically, one mood state may predominate over the other, or they may be mixed or combined with each other.

A bipolar person in the depressive phase may be sad, despondent, listless, lacking in energy, and unable to show interest in their surroundings or to enjoy themselves; they may also have a poor appetite and disturbed sleep. The depressive state can be agitated – in which case sustained tension, overactivity, despair, and apprehensive delusions predominate – or it can be retarded – in which case the person's activity is slowed and reduced, they are sad and dejected, and they suffer from self-depreciatory and self-condemnatory tendencies. Mania is a mood disturbance that is characterized by abnormally intense excitement, elation, expansiveness, boisterousness, talkativeness, distractibility, and irritability. The manic person talks loudly, rapidly, and continuously and progresses rapidly from one topic to another; is extremely enthusiastic, optimistic, and confident; is highly sociable and gregarious; gesticulates and moves about almost continuously; is easily irritated and easily distracted; is prone to grandiose notions; and shows an inflated sense of self-esteem. The most extreme manifestations of these two mood disturbances are, in the manic phase, violence against others and, in the depressive, suicide. A bipolar disorder may also feature such psychotic symptoms as delusions and hallucinations. Depression is the more common symptom, and many patients never develop a genuine manic phase, although they may experience a brief period of over-optimism and mild euphoria while recovering from a depression.

Bipolar disorders of varying severity affect about one per cent of the general population and account for 10 to 15 per cent of re-admissions to mental institutions. Statistical studies have suggested a hereditary predisposition to bipolar disorder, and this predisposition has now been linked to a defect on a dominant gene located on chromosome 11. In a physiological sense, it is believed that bipolar disorder is caused by the faulty regulation of one or more naturally occurring amines at sites in the brain where the transmission of nerve impulses takes place; a deficiency of the amines results in depression, and an excess of them causes mania. The most likely candidates for the suspect amines are norepinephrine, dopamine, and 5-hydroxytryptamine. The ingestion of lithium carbonate on a long-term basis has been found effective in alleviating or even eliminating the symptoms of many people with bipolar disorder.

Eating Disorders

Two of the more common eating disorders involve not only abnormalities of eating behaviour but also distortions in body perception. Anorexia nervosa consists of a considerable loss in body weight, refusal to gain weight, and a fear of becoming overweight that is dramatically at odds with reality. People with anorexia often become grotesquely thin in the eyes of everyone but themselves, and they manifest the physical symptoms of starvation. Bulimia nervosa is characterized by impulsive or "binge" eating, alternating with maladaptive (and ineffective) efforts to lose weight, such as by purging (e.g. vomiting or using laxatives) or fasting. People with bulimia are also preoccupied with body weight and shape, but they do not exhibit the weight loss apparent in anorexia patients.

Anorexia and bulimia are contrasting disorders with respect to self-control; those with anorexia apply excessive control over their eating behaviour, while those with bulimia exhibit a loss of control at some times with attempts to compensate for this at other times. The Diagnostic and Statisical Manual of Mental Disorders reports lifetime prevalence rates of 0.5 per cent for anorexia nervosa and between one and three per cent for bulimia nervosa. The typical age of onset for both disorders is mid- to late adolescence. The disorders are diagnosed far more frequently in girls than in boys.

Misperceptions of one's appearance can also be manifested as body dysmorphic disorder, in which an individual magnifies the negative aspects of a perceived flaw to such a degree that they shun social settings or embark compulsively upon a series of appearance-augmenting procedures, such as dermatological treatments and plastic surgery, in an attempt to remove the perceived defect.

As we have seen in the discussion of depression and the other disorders above, psychiatrists and psychologists treat neuroses in a variety of ways. The psychoanalytic approach involves helping the patient to become aware of the repressed impulses, feelings, and traumatic memories that underlie their symptoms, thereby enabling them to achieve personality growth through a better and deeper self-understanding. Those who hold that neuroses are the result of learned responses may recondition a neurotic patient through a process known as desensitization: a patient afraid of heights, for example, would be gradually exposed to progressively greater heights over several weeks. Other learning approaches include modelling more effective behaviour, wherein the patient learns by example. Cognitive and interpersonal approaches include discussing thoughts and perceptions that contribute to a patient's neurotic symptoms, eventually replacing them with more realistic interpretations of external events and the patient's internal responses to them. Many psychiatrists prefer physical approaches, such as psychotropic drugs (including anti-anxiety agents and antidepressant and antipsychotic drugs) and electroconvulsive (shock) therapy. Many psychiatrists advocate combinations of these approaches, the exact nature of which depend on the patient and their complaint.

Why it Happens

Very often the aetiology, or cause, of a particular type of mental disorder is unknown or is understood only to a very limited extent. The situation is complicated by the fact that a mental disorder such as schizophrenia may be caused by a combination and interaction of several factors, including a probable genetic predisposition to develop the disease, a postulated biochemical imbalance in the brain, and a cluster of stressful life events that help to precipitate the actual onset of the illness. The predominance of these and other factors probably varies from person to person in schizophrenia. A similarly complex interaction of constitutional, developmental, and social factors can influence the formation of mood and anxiety disorders.

No single theory of causation can explain all mental disorders or even all those of a particular type. Moreover, the same type of disorder may have different causes in different people, for example an OCD may have its origins in a biochemical imbalance, in an unconscious emotional conflict, in faulty learning processes, or in a combination of these. The fact that quite different therapeutic approaches can produce equal improvements in different patients with the same type of disorder underscores the complex and ambiguous nature of the causes of mental illness. The major theoretical and research approaches to the causation of mental disorders are treated below.

Genetics

The study of the genetic causes of mental disorders involves both the laboratory analysis of the human genome and the statistical analysis of the frequency of a particular disorder's occurrence among individuals who share related genes, i.e. family members and particularly twins. Family risk studies compare the observed frequency of occurrence of a mental illness in close relatives of the patient with its frequency in the general population. First-degree relatives (parents, siblings, and children) share 50 per cent of their genetic material with the patient, and higher rates of the illness in these relatives than expected indicate a possible genetic factor. In twin studies the frequency of occurrence of the illness in both members of pairs of identical (monozygous) twins is compared with its frequency in both members of a pair of fraternal (dizygous) twins. A higher concordance for disease among the identical than the fraternal twins suggests a genetic component. Further information on the relative importance of genetic and environmental factors accrues from comparing identical twins reared together with those reared apart. Adoption studies comparing adopted children whose biological parents had the illness with those whose parents did not can also be useful in separating biological from environmental influences.

Such studies have demonstrated a clear role for genetic factors in the causation of schizophrenia. When one parent is found to have the disorder, the probability of that person's children developing schizophrenia is at least 10 times higher (about a 12 per cent risk probability) than it is for children in the general population (about a one per cent risk probability). If both parents have schizophrenia, the probability of their children developing the disorder is anywhere from 35 to 65 per cent. If one member of a pair of fraternal twins develops schizophrenia, there is about a 12 per cent chance that the other twin will too. If one member of a pair of identical twins has schizophrenia, the other identical twin has at least a 40 to 50 per cent chance of developing the disorder. Although genetic factors seem to play a less significant role in the causation of other psychotic and personality disorders, studies have demonstrated a probable role for genetic factors in the causation of many mood disorders and some anxiety disorders.

Biochemistry

If a mental disease is caused by a biochemical abnormality, investigation of the brain at the site where the biochemical imbalance occurs should show neurochemical differences from normal. In practice such a simplistic approach is fraught with practical, methodological, and ethical difficulties. The living human brain is not readily accessible to direct investigation, and the dead brain undergoes chemical change; moreover, findings of abnormalities in cerebrospinal fluid, blood, or urine may have no relevance to the question of a presumed biochemical imbalance in the brain. It is difficult to study human mental illnesses using animals as analogues, because most mental disorders either do not occur or are not recognizable in animals. Even when biochemical abnormalities have been found in people with mental disorders, it is difficult to know whether they are the cause or the result of the illness, or of its treatment, or of other consequences. Despite these problems, progress has been made in unravelling the biochemistry of mood disorders, schizophrenia, and some of the dementias.

Certain drugs have been demonstrated to have beneficial effects upon mental illnesses. Antidepressant, antipsychotic, and antianxiety drugs are thought to achieve their therapeutic results by the selective inhibition or enhancement of the quantities, action, or breakdown of neurotransmitters in the brain. Neurotransmitters are a group of chemical agents that are released by neurons (nerve cells) to stimulate neighbouring neurons, thus allowing impulses to be passed from one cell to the next throughout the nervous system. Neurotransmitters play a key role in transmitting nerve impulses across the microscopic gap (synaptic cleft) that exists between neurons. The release of such neurotransmitters is stimulated by the electrical activity of the cell. Norepinephrine, dopamine, acetylcholine, and serotonin are among the principal neurotransmitters. Some neurotransmitters excite or activate neurons, while others act as inhibiting substances. Abnormally low or high concentrations of neurotransmitters at sites in the brain are thought to change the synaptic activities of neurons, thus ultimately leading to the disturbances of mood, emotion, or thought found in various mental disorders.

Neuropathology

In the past the post-mortem study of the brain revealed information upon which great advances in understanding the aetiology of neurological and some mental disorders were based, leading to the German psychiatrist Wilhelm Griesinger's postulate: "all mental illness is disease of the brain". The application of the principles of pathology to general paresis, one of the most common conditions found in mental hospitals in the late nineteenth century, resulted in the discovery that this was a form of neurosyphilis and was caused by infection with the spirochete bacterium *Treponema pallidum*. The examination of the brains of patients with other forms of dementia has given useful information concerning other causes of this syndrome – for example, Alzheimer's disease and arteriosclerosis. The pinpointing of abnormalities of specific areas of the brain has aided understanding of some abnormal mental functions, such as disturbances of memory and speech disorders. Recent advances in neuroimaging techniques have expanded the ability to investigate brain abnormalities in patients with a wide variety of mental illnesses, eliminating the need for post-mortem studies.

Personality Disorders

Personality disorders are mental disorders that are marked by deeply ingrained and lasting patterns of inflexible, maladaptive, or antisocial behaviour. A personality disorder is an accentuation of one or more personality traits to the point that the trait significantly impairs an individual's social or occupational functioning. Personality disorders are not, strictly speaking, illnesses, because they need not involve the disruption of emotional, intellectual, or perceptual functioning. In many cases, an individual with a personality disorder does not seek psychiatric treatment for such unless they are pressured to by relations or by a court.

There are many different types of personality disorders; they are classified according to the particular personality traits that are accentuated. People with a paranoid personality disorder show a pervasive and unjustified mistrust and suspiciousness of others. They may be secretive or aggressive and are excessively sensitive to implied slights or criticism. People with schizoid personality disorder appear aloof, withdrawn, unresponsive, humourless, and dull and are solitary to an abnormal degree. People with explosive personality disorder exhibit extreme emotional instability characterized by explosive outbursts of rage upon minor provocation. People with histrionic personality disorder persistently display overly dramatic, highly excitable, and intensely expressed behaviour (i.e. histrionics). People with dependent personality disorder lack energy and initiative and passively let others assume responsibility for major aspects of their lives. Those with passive–aggressive personality disorder express their hostility through such indirect means as stubbornness, procrastination, inefficiency, and forgetfulness.

One of the most important disorders is the antisocial, sociopathic, or psychopathic personality disorder. This disorder is chiefly characterized by a personal history of chronic and continuous antisocial behaviour in which the rights of others are violated. Poor or non-existent job performance is another major indicator. People with antisocial personality disorder make up a significant portion of the criminal and delinquent elements of society. Besides persistent criminality, the symptoms may also include sexual promiscuity or sexual aggression and drug addiction or alcoholism. Sociopaths generally accept their behaviour as natural, feel no guilt when they hurt others, see little reason for or possibility of change, and resist therapy.

The causes of personality disorders are unknown, though there is undoubtedly a hereditary element involved. Personality traits are, by definition, virtually permanent, and so personality disorders are only partially amenable to treatment, if at all. The most effective treatment combines various behavioural and psychotherapeutic therapies. Medication may be helpful in alleviating periodic anxiety, depression, emotional instability, or paranoid tendencies in some cases.

Theories of Personality Development

In the first half of the twentieth century, theories of the aetiology of mental disorders, especially of neuroses and personality disorders, were dominated in the USA by Freudian psychoanalysis as devised by Sigmund Freud, and the derivative theories of the post-Freudians. In Western Europe the influence of Freudian theory upon psychiatric theory diminished after the Second World War.

Freudian and other psychodynamic theories view neurotic symptoms as arising from intrapsychic conflict, i.e. the existence of conflicting motives, drives, impulses, and feelings held within various components of the mind. Central to psychoanalytic theory is the postulated existence of the unconscious, which is that part of the mind whose processes and functions are inaccessible to the individual's conscious awareness or scrutiny. One of the functions of the unconscious is thought to be that of a repository for traumatic memories, feelings, ideas, wishes, and drives that are threatening, abhorrent, anxiety-provoking, or socially or ethically unacceptable to the individual. These mental contents may at some time be pushed out of conscious awareness but remain actively held in the unconscious. This process is a defence mechanism for protecting the individual from the anxiety or other psychic pain associated with those contents and is known as repression. The repressed mental contents held in the unconscious retain much of the psychic energy or power that was originally attached to them, however, and they can continue to influence significantly the mental life of the individual even though (or because) someone is no longer aware of them.

The natural tendency for repressed drives or feelings, according to this theory, is to reach conscious awareness so that the individual can seek the gratification, fulfilment, or resolution of them. But this threatened release of forbidden impulses or memories provokes anxiety and is seen as threatening, and a variety of defence mechanisms may then come into play to provide relief from the state of psychic conflict. Through reaction formation, projection, regression, sublimation, rationalization, and other defence mechanisms, some component of the unwelcome mental contents can emerge into consciousness in a disguised or attenuated form, thus providing partial relief to the individual. Later, perhaps in adult life, some event or situation in the person's life triggers the abnormal discharge of the pent-up emotional energy in the form of neurotic symptoms in a manner mediated by defence mechanisms. Such symptoms can form the basis of neurotic disorders such as conversion and somatoform disorders, anxiety disorders, obsessional disorders, and depressive disorders (see above). Because the symptoms represent a compromise within the mind between letting the repressed mental contents out and continuing to deny all conscious knowledge of them, the particular character and aspects of an individual's symptoms and neurotic concerns bear an inner meaning that symbolically represents the underlying intrapsychic conflict. Psychoanalysis and other dynamic therapies help a person achieve a controlled and therapeutic recovery that is based on a conscious awareness of repressed mental conflicts along with an understanding of their influence on past history and present difficulties. These steps are associated with the relief of symptoms and improved mental functioning.

Freudian theory views childhood as the primary breeding ground of neurotic conflicts. This is because children are relatively helpless and are dependent on their parents for love, care, security, and support and because their psychosexual, aggressive, and other impulses are not yet integrated into a stable personality framework. The theory posits that children lack the resources to cope with emotional traumas, deprivations, and frustrations; if these develop into unresolved intrapsychic conflicts that the young person holds in abeyance through repression, there is an increased likelihood that insecurity, unease, or guilt will subtly influence the developing personality, thereby affecting the person's interests, attitudes, and ability to cope with later stresses.

After Freud

Psychoanalytic theory's emphasis on the unconscious mind and its influence on human behaviour resulted in a proliferation of other, related theories of causation incorporating – but not limited to – basic psychoanalytic precepts. Most subsequent psychotherapies have stressed in their theories of causation aspects of earlier, maladaptive psychological development that had been missed or under-emphasized by orthodox psychoanalysis, or they have incorporated insights taken from learning theory. Swiss psychiatrist Carl Jung, for instance, concentrated on the individual's need for spiritual development and concluded that neurotic symptoms could arise from a lack of self-fulfilment in this regard. Austrian psychiatrist Alfred Adler emphasized the importance of feelings of inferiority and the unsatisfactory attempts to compensate for it as important causes of neurosis. Neo-Freudian authorities such as Harry Stack Sullivan, Karen Horney, and Erich Fromm modified Freudian theory by emphasizing social relationships and cultural and environmental factors as being important in the formation of mental disorders.

More modern psychodynamic theories have moved away from the idea of explaining and treating neurosis on the basis of a defect in a single psychological system and have instead adopted a more complex notion of multiple causes, including emotional, psychosexual, social, cultural, and existential ones. A notable trend was the incorporation of approaches derived from theories of learning. Such psychotherapies emphasized the acquired, faulty mental processes and maladaptive behavioural responses that act to sustain neurotic symptoms, thereby directing interest toward the patient's extant circumstances and learned responses to those conditions as a causative factor in mental illness. These approaches marked a convergence of psychoanalytic theory and behavioural theory, especially with regard to each school's view of disease causation.

Behavioural Theory

Behavioural theories for the causation of mental disorders, especially neurotic symptoms, are based upon learning theory, which was in turn largely derived from the study of the behaviour of animals in laboratory settings. Most important theories in this area arose out of the work of the Russian physiologist Ivan Pavlov and several American psychologists, such as Edward L. Thorndike, Clark L. Hull, John B. Watson, Edward C. Tolman, and B.F. Skinner. In the classic Pavlovian model of conditioning, an unconditioned stimulus is followed by an appropriate response; for example, food placed in a dog's mouth is followed by the dog salivating. If a bell is rung just before food is offered to a dog, eventually the dog will salivate at the sound of the bell only, even though no food is offered. Because the bell could not originally evoke salivation in the dog (and hence was a neutral stimulus) but came to evoke salivation because it was repeatedly paired with the offering of food, it is called a conditioned stimulus. The dog's salivation at the sound of the bell alone is called a conditioned response. If the conditioned stimulus (the bell) is no longer paired with the unconditioned stimulus (the food), extinction of the conditioned response gradually occurs (the dog ceases to salivate at the sound of the bell alone).

Behavioural theories for the causation of mental disorders rest largely upon the assumption that the symptoms or symptomatic behaviour found in people with various neuroses (particularly phobias and other anxiety disorders) can be regarded as learned behaviours that have been built up into conditioned responses. In the case of phobias, for example, someone who has once been exposed to an inherently frightening situation afterward experiences anxiety even at neutral objects that were merely associated with that situation at the time but that should not reasonably produce anxiety. Thus, a child who has had a frightening experience with a bird may subsequently have a fear response to the sight of feathers. The neutral object alone is enough to arouse anxiety, and the person's subsequent effort to avoid that object is a learned behavioural response that is self-reinforcing, because the person does indeed procure a reduction of anxiety by avoiding the feared object and is thus likely to continue to avoid it in the future. It is only by confronting the object that the individual can eventually lose the irrational, association-based fear of it.

12

DRUGS AND THE BRAIN

Drug use is the use of drugs for psychotropic rather than medical purposes. Among the most common psychotropic drugs are opiates (opium, morphine, heroin), hallucinogens (LSD, mescaline, psilocybin), barbiturates, cocaine, amphetamines, tranquillizers, and cannabis. Alcohol and tobacco are also sometimes classified as drugs. The term "drug abuse" is normally applied to excessive and addictive use of drugs. Because such drugs can have severe physiological and psychological, as well as social, effects, many governments regulate their use.

To consider drugs only as medicinal agents or to insist that drugs be confined to prescribed medical practice is to fail to understand humans. The remarks of the American sociologist Bernard Barber are poignant in this regard:

> Not only can nearly anything be called a "drug", but things so called turn out to have an enormous variety of psychological and social functions – not only religious and therapeutic and "addictive", but political and aesthetic and ideological and aphrodisiac and so on. Indeed, this has

been the case since the beginning of human society. It seems that always and everywhere drugs have been involved in just about every psychological and social function there is, just as they are involved in every physiological function.

Opium, Morphine, Heroin, and Related Synthetics

The various opiates and related synthetics all produce about the same physiological effects. All are qualitatively similar to morphine in action and differ from each other mainly in degree. The most long-lasting and conspicuous physiological responses are obtained from the central nervous system and the smooth muscle of the gastrointestinal tract. These effects, while restricted, are complex and vary with the dosage and the route of administration (intravenous, subcutaneous, oral). Both depressant and stimulant effects are elicited. The depressant action involves the cerebral cortex, with a consequent narcosis, general depression, and reduction in pain perception; it also involves the hypothalamus and brainstem, inducing sedation, the medulla, with associated effects on respiration, the cough reflex, and the vomiting centre (late effect). The stimulant action involves the spinal cord and its reflexes, the vomiting centre (early effect), the tenth cranial nerve with a consequent slowing of the heart, and the third cranial nerve resulting in pupil constriction. Associated effects of these various actions include nausea, vomiting, constipation, itchiness of the facial region, yawning, sweating, flushing of skin, a warm sensation in the stomach, fall in body temperature, diminished respiration, and heaviness in the limbs.

The most outstanding effect of the opiates is one of analgesia. All types of pain perception are affected, but the best analgesic response is obtained in relieving dull pain. The analgesic effects increase with increasing doses until a limit is reached beyond which no further improvement is obtained. This point may fall just short of complete relief.

Depression of cortical function results in a euphoric response involving a reduction of fear and apprehension, a lessening of inhibitions, an expansion of ego, and an elevation of mood that combine to enhance the general sense of well-being. Occasionally in pain-free individuals, the opposite effect, dysphoria, occurs and there is anxiety, fear, and some depression. In addition to analgesia and associated euphoria, there is drowsiness, mental and physical impairment, a clouding of consciousness, poor concentration and attention, reduced hunger or sex drives, and sometimes apathy.

Apart from their addiction liability, respiratory depression leading to respiratory failure and death is the chief hazard of these drugs. All of the more potent opiates and synthetics produce rapid tolerance, and tolerance to one member of this group is always associated with tolerance to the other members of the group (cross-tolerance). The more potent members of the group have a very great addiction liability with the associated physical dependence and abstinence syndrome.

Opium

Opium is a narcotic drug that is obtained from the unripe seed pods of the opium poppy (*Papaver somniferum*), a plant of the family Papaveraceae. Opium is obtained by slightly incising the seed capsules of the poppy after the plant's flower petals have fallen. The slit seed pods exude a milky latex that coagulates and changes colour, turning into a gum-like brown mass upon exposure to air. This raw opium may be ground into a powder, sold as lumps, cakes, or bricks, or treated further to obtain such derivatives as morphine, codeine, and heroin. Opium and the drugs obtained from it are called opiates.

The pharmacologically active principles of opium reside in its alkaloids, the most important of which, morphine, constitutes about 10 per cent by weight of raw opium. Other active alkaloids such as papaverine and codeine are present in smaller proportions. Opium alkaloids are of two types, depending on chemical structure and action. Morphine, codeine, and thebaine, which represent one type, act upon the central nervous system and are analgesic, narcotic, and potentially addicting compounds. Papaverine, noscapine (formerly called narcotine), and most of the other opium alkaloids act only to relax involuntary (smooth) muscles.

Opiates exert their main effects on the brain and spinal cord. Their principal action is to relieve or suppress pain. The drugs also alleviate anxiety; induce relaxation, drowsiness, and sedation; and may impart a state of euphoria or other enhanced mood. Opiates also have important physiological effects; they slow respiration and heartbeat, suppress the cough reflex, and relax the smooth muscles of the gastrointestinal tract. Opiates are addictive drugs, that is, they produce a physical dependence (and withdrawal symptoms) that can only be assuaged by continued use of the drug. With chronic use, however, the body develops a tolerance to opiates, so that progressively larger doses are needed to achieve the same effect. The higher opiates – heroin and morphine – are more addictive than opium or codeine. Opiates are classified as narcotics because they relieve pain, induce stupor and sleep, and produce addiction. The habitual use of opium produces physical and mental deterioration and shortens life. An acute overdose of opium causes respiratory depression, which can be fatal.

Opium was for many centuries the principal painkiller known to medicine and was used in various forms and under various names. Laudanum, for example, was an alcoholic tincture (dilute solution) of opium that was used in European medical practice as an analgesic and sedative. Since the late 1930s, various synthetic drugs have been developed that possess the analgesic properties of morphine and heroin. These drugs, which include meperidine (Demerol), methadone, levorphonal, and many others, are known as synthetic opioids. They have largely replaced morphine and heroin in the treatment of severe pain.

Opiates achieve their effect on the brain because their structure closely resembles that of certain molecules called endorphins (see below), which are naturally produced in the body. Endorphins suppress pain and enhance mood by occupying certain receptor sites on specific neurons (nerve cells) that are involved in the transmission of nervous impulses. Opiate alkaloids are able to occupy the same receptor sites, thereby mimicking the effects of endorphins in suppressing the transmission of pain impulses within the nervous system.

Endorphins

An endorphin is any of a group of opiate proteins with pain-relieving properties that are found naturally in the brain. The main substances identified as endorphins include the enkephalins, beta-endorphin, and dynorphin, which were discovered in the 1970s by Roger Guillemin and other researchers. Endorphins are distributed in characteristic patterns throughout the nervous system, with beta-endorphin found almost entirely in the pituitary gland.

Endorphins have been found to be clearly involved in the regulation of pain; even the analgesic effects of acupuncture treatments may be attributable to them. Such substances are also believed to have some relation to appetite control, the release of sex hormones through the pituitary, and the adverse effects of shock. There is strong evidence that endorphins are connected with "pleasure centres" in the brain. Knowledge about the behaviour of the endorphins and their receptors in the brain has implications for the treatment of opiate addictions and chronic pain disorders.

Hallucinogens

A hallucinogen is a substance that produces psychological effects that are normally associated only with dreams, schizophrenia, or religious exaltation. It produces changes in perception, thought, and feeling, ranging from distortions of what is sensed (illusions) to sensing objects where none exist (hallucinations). Hallucinogens heighten sensory signals, but this is often accompanied by loss of control over what is experienced.

The psychopharmacological drugs that have aroused widespread interest and bitter controversy are those that produce marked aberrations in behaviour or perception. Among the most prevalent of these are D-lysergic acid diethylamide, or LSD-25, which originally was derived from ergot (*Claviceps purpurea*), a fungus on rye and wheat; mescaline, the active principle of the peyote cactus (*Lophophora williamsii*), which grows in the southwestern USA and Mexico; and psilocybin and psilocin, which come from certain mushrooms (notably two Mexican species, *Psilocybe mexicana* and *Stropharia cubensis*). Other hallucinogens include bufotenine, originally isolated from the skin of toads; harmine, from the seed coats of a plant of the Middle East and Mediterranean region; and the synthetic compounds methylenedioxyamphetamine (MDA), methylenedioxymethamphetamine (MDMA), and phencyclidine (PCP). Tetrahydrocannabinol (THC), the active ingredient in cannabis, or marijuana, obtained from the leaves and tops of the hemp plant (*Cannabis sativa*), is also sometimes classified as a hallucinogen.

Apparently, for thousands of years native societies of the Western hemisphere utilized plants containing psychedelic substances. The hallucinogenic mushrooms of Mexico were considered sacred and were called "god's flesh" by the Aztecs, and during the nineteenth century the Mescalero Apaches of the southwestern USA practised a peyote rite that was adopted by many of the Plains tribes. Peyotism eventually became fused with Christianity, and the Native American Church was formed in 1918 to protect peyotism as a form of worship.

Scientific interest in hallucinogens developed slowly. Mescaline was finally isolated as the active principle of peyote in 1896. It was not until 1943, when the Swiss chemist Albert Hofmann accidentally ingested a synthetic preparation of LSD and experienced its psychedelic effects, that the search for a natural substance responsible for schizophrenia became widespread. Gordon Wasson, a New York banker and mycologist, called attention to the powers of the Mexican mushrooms in 1953, and the active principle was quickly found to be psilocybin.

Only the D-isomer of LSD is found to be psychedelically active. It is thought that LSD, as well as psilocybin, psilocin, bufotenine, and harmine, act antagonistically toward serotonin, an important brain amine. However, evidence for this is quite contradictory. Some chemicals that block serotonin receptors in the brain have no psychedelic activity. Mescaline is structurally related to the adrenal hormones epinephrine and norepinephrine – catecholamines that are very active in the peripheral nervous system and are suspected of playing a role as neurotransmitters in the central nervous system.

During the 1950s and 1960s there was a great deal of scientific research with hallucinogens in psychotherapy. LSD was used in the treatment of alcoholism, to reduce the suffering of terminally ill cancer patients, and in the treatment of children with autism. Controversy arose over the social aspects of the drugs. Subsequent scientific research indicated that the side-effects of these drugs were more serious than previous research had indicated and that human experimentation was somewhat premature. As a result, many of the hallucinogens were limited to scientific use, with pharmaceutical manufacture strictly regulated.

Illicit experimentation continued over the following decades, partly inspired by the mystical writings of Aldous Huxley, and a vigorous subculture sprang up surrounding hallucinogens in the 1960s. Originating on the West Coast of the USA, it spread throughout North America, Western Europe, and Australia. At the end of the century there was a revived interest in LSD in the USA, and the drug Ecstasy became popular among young people. In addition, some individuals began experimenting with countless new substances, particularly from the phenethylamine and tryptamine families, which was difficult to regulate or suppress because the necessary information to make the drugs was widely available through the Internet.

Sedative-Hypnotic Drugs

A sedative-hypnotic drug is a chemical substance used to reduce tension and anxiety and induce calm (sedative effect) or to induce sleep (hypnotic effect). Most such drugs exert a quieting or calming effect at low doses and a sleep-inducing effect in larger doses. Sedative-hypnotic drugs tend to depress the central nervous system. Because these actions can be obtained with other drugs, such as opiates, the distinctive characteristic of sedative-hypnotics is their selective ability to achieve their effects without affecting mood or reducing sensitivity to pain.

For centuries alcohol and opium were the only drugs available that had sedative-hypnotic effects. Chloral hydrate, a derivative of ethyl alcohol, was introduced in 1869 as the first synthetic sedative-hypnotic, and a more important drug, barbital, was synthesized in 1903. Phenobarbital became available in 1912 and was followed, during the next 20 years, by a long series of other barbiturates. In the mid-twentieth century new types of sedative-hypnotic drugs were synthesized, chief among them the benzodiazepines (the so-called minor tranquillizers).

Barbiturates were extensively used as "sleeping pills" throughout the first half of the twentieth century. Among the most commonly prescribed kinds were phenobarbital, secobarbital (marketed under Seconal and other trade names), amobarbital (Amytal), and pentobarbital (Nembutal). When taken in high enough doses, these drugs are capable of producing a deep unconsciousness that makes them useful as general anaesthetics. In still higher doses, however, they depress the central nervous and respiratory systems to the point of coma, respiratory failure, and death. Additionally, the prolonged use of barbiturates for relief of insomnia leads to tolerance, in which the user requires amounts of the drug much in excess of the initial therapeutic dose, and to addiction, in which denial of the drug precipitates withdrawal, as indicated by such symptoms as restlessness, anxiety, weakness, insomnia, nausea, and convulsions.

Because of these health risks, the barbiturates were gradually supplanted by the benzodiazepines beginning in the 1960s. The latter are more effective in relieving anxiety than in inducing sleep, but they are superior to barbiturates because of the reduced dangers they present of tolerance and addiction and because they are much less likely to injuriously depress the central nervous system when used at high doses. They also require a much smaller dosage than do barbiturates to achieve their effects. The benzodiazepines include chlordiazepoxide (Librium), diazepam (Valium), alprazolam (Xanax), oxazepam (Serax, Serenid), and triazolam (Halcion). They are, however, intended only for short- or medium-term use, because the body does develop a tolerance to them and withdrawal symptoms (anxiety, restlessness, and so on) develop even in those who have used the drugs for only four to six weeks. The benzodiazepines are thought to accomplish their effect within the brain by facilitating the action of the neurotransmitter gamma-aminobutyric acid, which is known to inhibit anxiety.

Antipsychotic drugs (major tranquillizers), tricyclic antidepressants, and antihistamines can also induce drowsiness, though this is not their primary function. Most over-the-counter sleeping aids use antihistamines as their active ingredient.

Stimulants

A stimulant is any drug that excites any bodily function, but more specifically those that stimulate the brain and central nervous system. Stimulants induce alertness, elevated mood, wakefulness, increased speech and motor activity and decrease appetite. Their therapeutic use is limited, but their mood-elevating effects make some of them potent drugs of abuse.

The major stimulant drugs are amphetamines and related compounds, methylxanthines (methylated purines), cocaine, and nicotine. Amphetamines achieve their effect by increasing the amount and activity of the neurotransmitter norepinephrine (noradrenaline) within the brain. They facilitate the release of norepinephrine by nerve cells and interfere with the cells' reuptake and breakdown of the chemical, thereby increasing its availability within the brain. The most commonly used amphetamines are methamphetamine (Methedrine), amphetamine sulfate (Benzedrine), and dextroamphetamine sulfate (Dexamprex, Dexedrine). Amphetamines were first used in the 1930s to treat narcolepsy and subsequently became prescribed for obesity and fatigue. Their heavy or prolonged use causes irritability, restlessness, hyperactivity, anxiety, excessive speech, and rapid mood swings. Still higher doses or chronic use can cause agitation, tremor, confusion, and, in the most serious cases, a state resembling paranoid schizophrenia. Moreover, letdown effects of deep depression and physical exhaustion may occur after only a single dose of moderate strength wears off. With repeated use, tolerance develops, so that a user needs to take larger doses, but the accompanying dependence is not strong enough to be termed a physical addiction. Amphetamines are widely abused, in some cases by workers or students seeking enhanced physical energy and mental acuity to fulfil demanding tasks.

Certain drugs related to the amphetamines have the same mode of action but are somewhat milder stimulants. Among them are phenmetrazine (Preludin) and methylphenidate (Ritalin). The latter drug is widely used to "slow down" hyperactive children and improve their ability to concentrate.

The methylxanthines are even milder stimulants. Unlike the amphetamines and methylphenidate, which are synthetically manufactured, these compounds occur naturally in various plants and have been used by humans for many centuries. The most important of them are caffeine, theophylline, and theobromine. The strongest is caffeine, the active ingredient of coffee, tea, cola beverages, and maté. Theobromine is the active ingredient in cocoa. Caffeine constricts blood vessels of the brain; for this reason it is often a component of headache remedies. Theophylline is used in the treatment of severe asthma because of its capacity for relaxing the bronchioles in the lungs.

Cocaine is one of the strongest and shortest-acting stimulants and has a high potential for abuse owing to its euphoric and habit-forming effects. Nicotine, the active ingredient in cigarettes and other tobacco products, may also be regarded as a stimulant.

Adrenergic Drugs

The release of norepinephrine (noradrenaline) can be evoked or inhibited by the actions of adrenergic drugs. Drugs that evoke norepinephrine produce effects resembling those of sympathetic nerve activity and are called sympathomimetic agents. They include amphetamine and ephedrine, which act indirectly, mainly by expelling norepinephrine from its storage area in nerve terminals. They cause an increase in the heart rate (sometimes leading to arrhythmias, or irregular heartbeats) and other sympathetic effects. Ephedrine is occasionally used as a nasal decongestant. Amphetamine-like drugs also have strong effects on the brain, causing feelings of excitement and euphoria as well as reducing appetite, the latter effect leading to their use in treating obesity. Their effects on the brain have led to their recreational use and to their use as agents to enhance athletic performance. These drugs are liable to cause addiction, and overdosage may have dangerous cardiovascular and mental effects. Methylphenidate, an amphetamine-like compound sold under the trade name Ritalin, has been shown to be useful in the treatment of ADHD.

Effects of Alcohol on the Brain

Alcohol is a drug that affects the central nervous system. It belongs in a class with the barbiturates, minor tranquillizers, and general anaesthetics, and it is commonly classified as a depressant. The effect of alcohol on the brain is rather paradoxical. Under some behavioural conditions alcohol can serve as an excitant, under other conditions as a sedative. At very high concentrations it acts increasingly as a depressant, leading to sedation, stupor, and coma. The excitement phase exhibits the well-known signs of exhilaration, loss of socially expected restraints, loquaciousness, unexpected changes of mood, and unmodulated anger. Excitement actually may be caused indirectly, more by the effect of alcohol in suppressing inhibitory centres of the brain than by a direct stimulation of the manifested behaviour. The physical signs of excited intoxication are slurred speech, unsteady gait, disturbed sensory perceptions, and inability to make fine motor movements. Again, these effects are produced not by the direct action of alcohol on the misbehaving muscles and senses but by its effect on the brain centres that control the muscle activity.

The most important immediate action of alcohol is on the higher functions of the brain – those of thinking, learning, remembering, and making judgements. Many of the alleged salutary effects of alcohol on performance (such as better dancing, happier moods, sounder sleeping, less sexual inhibition, and greater creativity) have been shown in controlled experiments to be a function of suggestion and subjective assessment. In reality, alcohol improves performance only through muscle relaxation and guilt reduction or loss of social inhibition. Thus, mild intoxication actually makes objectively observed depression (and dancing for that matter) worse. Experiments also indicate a dependence of learning on the mental state in which it occurs. For example, what is learned under the influence of alcohol is better recalled under the influence of alcohol, but what is learned in the sober state is better recalled when sober.

Psychopharmacology

Psychopharmacology is the development, study, and use of drugs for the modification of behaviour and the alleviation of symptoms, particularly in the treatment of mental disorders. One of the most striking advances in the treatment of mental illnesses in the middle of the 20th century was the development of the series of pharmacological agents commonly known as tranquillizers (e.g. chlorpromazine, reserpine, and other milder agents) and antidepressants, including the highly effective group known as tricyclic antidepressants. Lithium is widely used to allay the symptoms of affective disorders and especially to prevent recurrences of both the manic and the depressed episodes in manic-depressive individuals. The many commercially marketed antipsychotic agents (including thiothixene, chlorpromazine, haloperidol, and thioridazine) all share the common property of blocking the dopamine receptors in the brain. (Dopamine acts to help transmit nerve impulses in the brain.) Since scientists have found a direct relationship between dopamine blockage and reduction of schizophrenic symptoms, many believe that schizophrenia may be related to excess dopamine.

These drugs contrast sharply with the hypnotic and sedative drugs that formerly were in use and that clouded the patient's consciousness and impaired their motor and perceptual abilities. The antipsychotic drugs can allay the symptoms of anxiety and reduce agitation, delusions, and hallucinations, and the antidepressants lift spirits and quell suicidal impulses. The heavy prescription use of drugs to reduce agitation and quell anxiety has led, however, to what many psychiatrists consider an overuse of such medications. An overdose of a tranquillizer may cause loss of muscular coordination and slowing of reflexes, and prolonged use can lead to addiction.

Toxic side-effects such as jaundice, psychoses, dependency, or a reaction similar to Parkinson's disease may develop. The drugs may produce other minor symptoms (e.g. heart palpitations, rapid pulse, sweating) because of their action on the autonomic nervous system.

Though particular drugs are prescribed for specific symptoms or syndromes, they are usually not specific to the treatment of any single mental disorder. Because of their ability to modify the behaviour of even the most disturbed patients, the antipsychotic, antianxiety, and antidepressant agents have greatly affected the management of the hospitalized mentally ill, enabling hospital staff to devote more of their attention to therapeutic efforts and enabling many patients to lead relatively normal lives outside of the hospital.

Behaviour and emotions are higher functional properties of the brain that depend on the network of neurons and chemical neurotransmitters that exist throughout the body; however, the means by which neurons achieve changes in behaviour and in mood remains unknown. Nevertheless, certain neurotransmitters, such as norepinephrine, dopamine, epinephrine, serotonin, and acetylcholine, appear to be closely linked to these aspects of brain function. Drugs that influence the operation of these neurotransmitter systems can profoundly influence and alter the behaviour of patients with mental disorders.

Psychiatric drugs, those that affect mood and behaviour, can be classified as follows: antianxiety agents, antidepressants, antimanics and antipsychotics (see below). Such drugs should be reserved for severe disruptions of normal emotional well-being and should not be used to relieve the boredom, tension, or sadness that may be properly regarded as a normal part of life.

Antipsychotic Drugs

Antipsychotic medications, which are also known as neuroleptics and major tranquillizers, belong to several different chemical groups but are similar in their therapeutic effects. These medications have a calming effect that is valuable in the relief of agitation, excitement, and violent behaviour in people with psychoses. The drugs are quite successful in reducing the symptoms of schizophrenia, mania, and delirium, and they are used in combination with antidepressants to treat psychotic depression. The drugs suppress hallucinations and delusions, alleviate disordered or disorganized thinking, improve the patient's lucidity, and generally make an individual more receptive to psychotherapy. Patients who have previously been agitated, intractable, or grossly delusional become noticeably calmer, quieter, and more rational when maintained on these drugs. The medications have enabled many patients with episodic psychoses to have shorter stays in hospitals and have allowed many other patients who would have been permanently confined to institutions to live in the outside world. The antipsychotics differ in their unwanted effects: some are more likely to make the patient drowsy; some to alter blood pressure or heart rate; and some to cause tremor or slowness of movement.

In the treatment of schizophrenia, antipsychotic drugs partially or completely control such symptoms as delusions and hallucinations. They also protect the patient who has recovered from an acute episode of the mental illness from suffering a relapse. The newer antipsychotic medications also treat social withdrawal, apathy, blunted emotional capacity, and the other psychological deficits characteristic of the chronic stage of the illness.

No single drug seems to be outstanding in the treatment of schizophrenia. In an individual patient, one drug may be preferred to another because it produces less severe unwanted effects, and the dose of any one drug needed to produce a therapeutic effect varies widely from patient to patient. Because of these individual differences, it is common for psychiatrists to substitute a drug of a different chemical group when one drug has been shown to be ineffective despite its use in adequate dosage for several weeks.

In an acute psychotic episode, a drug such as chlorpromazine, olanzepine, or haloperidol usually has a calming effect within a day or two. The control of psychotic symptoms such as hallucinations or disordered thinking may take weeks. The appropriate dosage has to be determined for each patient by cautiously increasing the dose until a therapeutic effect is achieved without unacceptable side-effects.

It is not known exactly how antipsychotic medications work. One theory is that they block dopamine receptors in the brain. Dopamine is a neurotransmitter, i.e. a chemical messenger produced by certain nerve cells that influence the function of other nerve cells by interacting with receptors in their cell membranes. Because schizophrenia may be caused by either the excessive release of or an increased sensitivity to dopamine in the brain, the effects of antipsychotic drugs may be due to their ability to block or inhibit dopamine transmission.

Dopamine-receptor blockade is certainly responsible for the main side-effects of first-generation antipsychotic medications. These symptoms, known as extrapyramidal symptoms (EPS), resemble those of Parkinson's disease and include tremor of the limbs; bradykinesia (slowness of movement with loss of facial expression, absence of arm-swinging during walking, and a general muscular rigidity); dystonia (sudden, sustained contraction of muscle groups causing abnormal postures); akathisia (a subjective feeling of restlessness leading to an inability to keep still); and tardive dyskinesia (involuntary movements, particularly involving the lips and tongue). Most extrapyramidal symptoms disappear when the drug is withdrawn. Tardive dyskinesia occurs late in the drug treatment and in about half of the cases persists even after the drug is no longer used. There is no satisfactory treatment for severe tardive dyskinesia.

Antianxiety Agents

The drugs most commonly used in the treatment of anxiety are the benzodiazepines, which have replaced the barbiturates because of their vastly greater safety. Benzodiazepines differ from each other in duration of action rather than effectiveness. Smaller doses have a calming effect and alleviate both the physical and psychological symptoms of anxiety. Larger doses induce sleep, and some benzodiazepines are marketed as hypnotics. The benzodiazepines have become among the most widely prescribed drugs in the developed world, and controversy has arisen over their excessive use by the public.

The side-effects of these medications are usually few – most often drowsiness and unsteadiness. Benzodiazepines are not lethal even in very large overdoses, but they increase the sedative effects of alcohol and other drugs. The benzodiazepines are basically intended for short- or medium-term use, because the body develops a tolerance to them that reduces their effectiveness and necessitates the use of progressively larger doses. Dependence on them may also occur, even in moderate dosages, and withdrawal symptoms have been observed in those who have used the drugs for only four to six weeks. In patients who have taken a benzodiazepine for many months or longer, withdrawal symptoms occur in 15 to 40 per cent of the cases and may take weeks or months to subside.

Withdrawal symptoms from benzodiazepines are of three kinds. Such severe symptoms as delirium or convulsions are rare. Frequently the symptoms involve a renewal or increase of the anxiety itself. Many patients also experience other symptoms such as hypersensitivity to noise and light as well as muscle twitching. As a result, many long-term users continue to take the drug not because of persistent anxiety but because the withdrawal symptoms are too unpleasant.

Because of the danger of dependence, benzodiazepines should be taken in the lowest possible dose for no more than a few weeks. For longer periods they should be taken intermittently, and only when the anxiety is severe.

Benzodiazepines act on specialized receptors in the brain that are adjacent to receptors for a neurotransmitter called gamma-aminobutyric acid (GABA), which inhibits anxiety. It is possible that the interaction of benzodiazepines with these receptors facilitates the inhibitory (anxiety-suppressing) action of GABA within the brain.

Antidepressant Agents

Many people suffering from depression gain symptomatic relief from treatment with an antidepressant. There are several classes of antidepressant drugs, which vary in their mechanism of action and side-effects. Successful treatment with such drugs relieves all the symptoms of depression, including disturbances of sleep and appetite, loss of sexual desire, and decreased energy, interest, and concentration. It usually takes two to three weeks for an antidepressant to improve someone's depressed mood significantly. Once a good response has been achieved, the drug should be continued for a further six months to reduce the risk of relapse. Antidepressants are also effective in treating other mental disorders such as panic disorder, agoraphobia, OCD, and bulimia nervosa.

It is widely theorized that depression is partly caused by reduced quantities or reduced activity of one or more neurotransmitters in the brain. Selective serotonin reuptake inhibitors (SSRIs), which include fluoxetine (Prozac) and sertraline (Zoloft, Lustral), are thought to act by inhibiting the reabsorption of the neurotransmitter serotonin. As a result, there is an accumulation of serotonin in the brain, a change that may be important in elevating mood. Because SSRIs interfere with only one neurotransmitter system, they have fewer, and less severe, side-effects than other classes of antidepressants, which inhibit the action of several neurotransmitters. Common side-effects of SSRIs include decreased sexual drive or ability, diarrhoea, insomnia, headache, and nausea.

Tricyclic antidepressants interfere with the reuptake of norepinephrine, serotonin, and dopamine. The side-effects of these drugs are mostly due to their interference with the function of the autonomic nervous system and may include dryness of the mouth, blurred vision, constipation, and difficulty urinating. Weight gain can be a distressing side-effect in people taking a tricyclic for a long period of time. In the elderly these drugs can cause delirium. Certain tricyclics interfere with conduction in heart muscle, and so they are best avoided in individuals with heart disease. Drug interactions occur with tricyclics, the most important being their interference with the action of certain drugs used in the treatment of high blood pressure.

Monoamine oxidase inhibitors (MAOIs) interfere with the action of monoamine oxidase, an enzyme involved in the breakdown of norepinephrine and serotonin. As a result, these neurotransmitters accumulate within nerve cells and presumably leak out on to receptors. The side-effects of these drugs include daytime drowsiness, insomnia, and a fall in blood pressure when changing position. The MAOIs interact dangerously with various other drugs, including narcotics and some over-the-counter drugs used in treating colds. Those taking an

MAOI must avoid certain foods containing tyramine or other naturally-occurring amines, which can cause a severe rise in blood pressure leading to headaches and even to stroke. Tyramine occurs in cheese, Chianti and other red wines, well-cured meats, and foods that contain monosodium glutamate (MSG).

Newer antidepressants, such as buproprion (Wellbutrin), have recently been introduced. These drugs are chemically unrelated to the other classes of antidepressants.

Mood-Stabilizing Drugs

Lithium, usually administered as its carbonate in several small doses per day, is effective in the treatment of an episode of mania. It can drastically reduce the elation, overexcitement, grandiosity, paranoia, irritability, and flights of ideas typical of people in the manic state. It has little or no effect for several days, however, and a therapeutic dose is rather close to a toxic dose. In severe episodes antipsychotic drugs may also be used. Lithium also has an antidepressant action in some patients with melancholia.

The most important use of lithium is in the treatment of patients with bipolar disorder or with recurrent depression. When given while the patient is well, lithium may prevent further mood swings, or it may reduce either their frequency or their severity. Its mode of action is unknown. Treatment begins with a small dose that is gradually increased until a specified concentration of lithium in the blood is reached. Blood tests to determine this are carried out weekly in the early stages of treatment and later every two to three months. It may take as long as a year for lithium to become fully effective.

The toxic effects of lithium, which usually occur when there are high concentrations of it in the blood, include drowsiness, coarse tremors, vomiting, diarrhoea, uncoordinated movement, and, with still higher blood concentrations, convulsions, coma, and death. At therapeutic blood concentrations, lithium's side-effects include fine tremors (which can be alleviated by propranolol), weight gain, passing increased amounts of urine with consequent increased thirst, and reduced thyroid function.

Carbamazepine, an anticonvulsant drug, has been shown to be effective in the treatment of mania and in the maintenance treatment of bipolar disorder. It may be combined with lithium in patients with bipolar disorder who fail to respond to either drug alone. Divalproex, another anticonvulsant, is also used in the treatment of mania.

Electroconvulsive Therapy

In electroconvulsive therapy (ECT), also called shock therapy, a seizure is induced in a patient by passing a mild electric current through the brain. The mode of action of ECT is not understood. Several studies have shown that ECT is effective in treating patients with severe depression, acute mania, and some types of schizophrenia. However, the procedure remains controversial and is used only if all other methods of treatment have failed.

Before administering ECT, the patient is given an intravenous injection of an anaesthetic in order to induce sleep and then is administered an injection of a muscle relaxant in order to reduce muscular contractions during the treatment. The electrical current is then applied to the brain. In bilateral ECT this is done by applying an electrode to each side of the head; in unilateral ECT both electrodes are placed over the non-dominant cerebral hemisphere, i.e. the right side of the head in a right-handed person. Unilateral ECT produces noticeably less confusion and memory impairment in patients, but more treatments may be needed. Patients recover consciousness rapidly after the treatment but may be confused and may experience a mild headache for an hour or two.

ECT treatments are normally given two or three times a week in the treatment of patients with depression. The number of electroconvulsive treatments required to treat depression is usually between six and 12. Some patients improve after the first treatment, others only after several. Once a program of ECT has been successfully completed, maintenance treatment with an antidepressant significantly decreases the patient's risk of relapse.

ECT is often considered for cases of severe depression when the patient's life is endangered because of refusal of food and fluids or because of serious risk of suicide, as well as in cases of postpartum depression, when it is desirable to reunite the mother and baby as soon as possible. ECT is often used in treating patients whose depression has not responded to adequate dosages of antidepressants.

The chief unwanted effect of ECT is impairment of memory. Some patients report memory gaps covering the period just before treatment, but others lose memories from several months before treatment. Many patients have memory difficulties for a few days or even a few weeks after completion of the treatment so that they forget appointments, phone numbers, and the like. These difficulties are transient and disappear rapidly in the vast majority of patients. Occasionally, however, patients complain of permanent memory impairment after ECT.

Psychosurgery

Psychosurgery is the destruction of groups of nerve cells or nerve fibres in the brain by surgical techniques in an attempt to relieve severe psychiatric symptoms. The removal of a brain tumour that is causing psychiatric symptoms is not an example of psychosurgery.

The classic technique of bilateral prefrontal leucotomy (lobotomy) is no longer performed because of its frequent undesirable effects on physical and mental health, in particular the development of epilepsy and the appearance of permanent, undesirable changes in personality. The latter include increased apathy and passivity, lack of initiative, and a generally decreased depth and intensity of the person's emotional responses to life. The procedure was used to treat chronically self-destructive, delusional, agitated, or violent psychotic patients. Stereotaxic surgical techniques have been developed that enable the surgeon to insert metal probes in specific parts of the brain; small areas of nerve cells or fibres are then destroyed by the implantation of a radioactive substance (usually yttrium) or by the application of heat or cold.

Proponents of psychosurgery claim that it is effective in treating some patients with severe and intractable OCD and that it may improve the behaviour of abnormally aggressive patients. However, many of the therapeutic effects that were claimed for psychosurgery by its adherents are now attainable by the use of antipsychotic and antidepressant medications. Today psychosurgery has a very small part to play in psychiatric treatment when the prolonged use of other forms of treatment has been unsuccessful and the patient is chronically and severely distressed or tormented by psychiatric symptoms. Whereas ECT is a routine treatment in certain specified conditions, psychosurgery is, at best, a last resort.

13

THE AGEING BRAIN

Physiologists have found that the performance of many organs such as the heart, kidneys, brain, or lungs shows a gradual decline over the life span. Part of this decline is due to a loss of cells from these organs, with resultant reduction in the reserve capacities of the individual. Furthermore, the cells remaining in the elderly individual may not perform as well as those in the young. Certain cellular enzymes may be less active, and thus more time may be required to carry out chemical reactions. Ultimately the cell may die.

Changes in the structures of the brain due to normal ageing are not striking. It is true that with advancing age there is a slight loss of neurons (nerve cells) in the brain. This is because, in the adult, neurons have lost the capacity to form new neurons by division. The basic number of neurons in the brain appears to be fixed by about the age of 10. The total number of neurons is extremely large, however, so that any losses probably have only a minor effect on behaviour. Because the physiological basis of memory is still unknown, it cannot be assumed that the loss of memory observed in elderly people is caused by the loss of neurons in the brain.

Neurons are extremely sensitive to oxygen deficiency. Consequently, it is probable that neuron loss, as well as other abnormalities observed in ageing brains, results not from ageing itself, but from disease, such as arteriosclerosis, that reduces the oxygen available to areas of the brain by reducing the blood supply.

There are probably functional changes in the brain that account for the slowing of responses and for the memory defects that are often seen in the elderly; and even small changes in the connections between cells of the brain could serve as the basis for marked behavioural changes, but, until more is known about how the brain works, behavioural changes cannot be related to physiological or structural changes. It is known that, because of the slow course of ageing, the nervous system can compensate and maintain adequate function even in centenarians.

Human behaviour is highly dependent on the reception and integration of information derived from sensory organs, such as the eye and ear, as well as from nerve endings in skin, muscle, joints, and internal organs. There is, however, no direct relation between the sensitivity of receptors and the adequacy of behaviour, because the usual level of stimulation is considerably greater than the minimum required for stimulation of the sense organs. In addition, an individual adapts to gradual impairments in one sensory organ by using information available from other sense organs. Modern technology has also provided glasses and hearing aids to compensate for reduced acuity in the sense organs.

The incidence of gross sensory impairments, of which many are the result of disease processes, increases with age. One survey conducted in the USA classified 25.9 per 1,000 people aged 65–74 as blind, in contrast to 1.3 per 1,000 aged 20–44. In the age group 65–74, 54.7 per 1,000 people were classified as functionally deaf, compared with 5.0 per 1,000 in the 25–34 age range.

The loss of psychological and neurophysiological capacities with age is undoubtedly the result, in large part, of the loss of neurons, but deficiencies in the metabolic processes of the surviving cells are demonstrably involved. The ability of the eye to dark-adapt (i.e. increase its sensitivity at low light levels) decreases with age, but part of that decrease can be restored by breathing pure oxygen. Various mental processes in old people are also found to be improved by breathing oxygen. The establishment of a memory trace (connections in the brain that are associated with memory) involves the synthesis of protein; any slowed induction of protein synthesis, as from lower oxygen intake, with age could be a factor in the deficits of learning and memory of old people.

A general characteristic of ageing of the endocrine system is that the cells that once responded vigorously to hormones become less responsive. A normal chemical in cells, cyclic adenosine monophosphate (AMP), is thought to be a transmitter of hormonal information across cell membranes; it may be possible to identify the specific sites in the membrane or the cell interior at which communication breaks down.

The most outstanding psychological features of ageing are the impairment in short-term memory and the lengthening of response time. Both of these factors contribute to lower scores of the elderly on standard tests of "intelligence". When the aged are given all the time that they wish on tests that are not heavily dependent on school skills, their performance is only slightly poorer than that of young adults. Age decrements are negligible on tests that depend on vocabulary, general information and well-practised activities.

Experimental studies on learning show that, although the elderly learn more slowly than the young, they can acquire new material and can remember it as well as the young. Age differences in learning increase with the difficulty of the material to be learned.

Aged people tend to become more cautious and rigid in their behaviour and to withdraw from social contacts. These behaviour patterns may be the result of social institutions and expectancies rather than an intrinsic phenomenon of ageing. Many people who "age successfully" make conscious efforts to maintain mental alertness by continued learning and by expansion of social contacts with individuals in a younger age group.

What Happens as the Brain Ages

A study published in the journal *Nature* raised considerable hope about the possibility of preventing Alzheimer's disease with a vaccine. Researchers at the San Francisco firm Elan Pharmaceuticals vaccinated mice that had been genetically programmed to overproduce amyloid, a protein–carbohydrate complex that forms harmful deposits in the brain, known as plaques. Amyloid plaques are a hallmark of Alzheimer's disease. In young healthy mice the vaccine prevented the formation of brain-clogging plaques altogether, and in older mice it prevented further progression of existing plaques.

Atrophy of brain or spinal cord tissue may be brought about by injuries that directly affect a localized area or that interfere with the blood supply to an area. When peripheral nerves are severed, degenerative and eventually atrophic changes ensue in the part beyond the injury. This type of atrophy is known as Wallerian degeneration. If conditions do not allow regeneration of nerve fibres from the proximal fragment of the cut nerve, atrophy is the eventual fate of the nerve tissue distal to the injury. Retrograde atrophy also occurs from disuse and affects the ganglion cells of the injured nerve.

Prolonged pressure brings about atrophy in the central nervous system as elsewhere. The pressure of an expanding tumour of the membranes covering the brain results in localized atrophy of the adjacent brain substance on which it impinges. In hydrocephalus more widespread atrophy of brain tissue results from the abnormal amounts of fluid confined within the rigid bony compartment of the skull. Increased pressure within the skull may force a portion of the brain through the foramen magnum, the bony opening at the base of the skull, and, if prolonged, results in a localized atrophy of cerebellar tissue pressed against the bony wall.

The late stages of chronic infections may be characterized by atrophy of the brain. A striking example of this is the variety of syphilitic infection of the nervous system known as general paresis in which the brain is shrunk and reduced in weight, the atrophy affecting mainly the cortex of the brain, particularly or most markedly in the frontal area. Occasionally the atrophy is local or affects only one side of the brain. The shrinkage of the brain tissue is mainly due to loss of many nerve cells of the cortex.

Death

Many dictionaries define death as "the extinction or cessation of life" or as "ceasing to be". As life itself is notoriously difficult to define – and as everyone tends to think of things in terms of what is known – the problems in defining death are immediately apparent. The most useful definitions of life are those that stress function, whether at the level of physiology, of molecular biology and biochemistry, or of genetic potential.

Death should be thought of as the irreversible loss of such functions.

The remainder of this chapter first explores the recurrent problems involved in seeking a biological definition of death. It then examines the implications of these problems in relation to human death. In this context, there are two major points: (1) death of the brain is the necessary and sufficient condition for death of the individual; and (2) the physiological core of brain death is the death of the brainstem. Finally, the chapter examines notions about the meaning of human death that have prevailed throughout history in a wide variety of cultural contexts. By so doing, it attempts to show that brainstem death, far from being a radically new idea, turns out to have always provided both an ultimate mechanism of death and a satisfactory anatomical basis for a wide range of philosophical concepts relating to death.

Brain death is physiological decapitation: it arises when intracranial pressure exceeds arterial pressure, thereby depriving the brain of its blood supply as efficiently as if the head had been cut off.

Brain death is synonymous with brainstem death, because the control centres for essential functions such as consciousness, respiration, and blood pressure are located within the brainstem. In many countries strict criteria for diagnosis of brain death have been established by common consent among medical, religious, ethical, and legal experts. Signs of brain death include the presence of deep coma with an established cause, the absence of any brainstem functions such as spontaneous respiration, pupillary reactions, eye movements, and gag and cough reflexes. Electroencephalography may be a useful confirmatory test. When brainstem death is confirmed, the heart usually stops beating within a day or two, even when other vital functions are artificially maintained.

The brainstem is the area at the base of the brain that includes the mesencephalon (midbrain), the pons, and the medulla. It contains the respiratory and vasomotor centres, which are responsible, respectively, for breathing and the maintenance of blood pressure. Most importantly, it also contains the ascending reticular activating system, which plays a crucial role in maintaining alertness (i.e. in generating the capacity for consciousness); small, strategically situated lesions in the midbrain and pons cause permanent coma. All of the motor outputs from the cerebral hemispheres – for example, those that mediate movement or speech – are routed through the brainstem, as are the sympathetic and parasympathetic efferent nerve fibres responsible for the integrated functioning of the organism as a whole. Most sensory inputs also travel through the brainstem. This part of the brain is, in fact, so tightly packed with important structures that small lesions there often have devastating effects. By testing various brainstem reflexes, moreover, the functions of the brainstem can be assessed clinically with an ease, thoroughness, and degree of detail not possible for any other part of the central nervous system.

It must be stressed that the capacity for consciousness (an upper brainstem function) is not the same as the content of consciousness (a function of the cerebral hemispheres); it is, rather, an essential precondition of the latter. If there is no functioning brainstem, there can be no meaningful or integrated activity of the cerebral hemispheres, no cognitive or affective life, no thoughts or feelings, no social interaction with the environment, nothing that might legitimize adding the adjective *sapiens* ("wise") to the noun *Homo* ("man"). The "capacity for consciousness" is perhaps the nearest one can get to giving a biological flavour to the notion of "soul".

The capacity to breathe is also a brainstem function, and apnea (respiratory paralysis) is a crucial manifestation of a non-functioning lower brainstem. Alone, of course, it does not imply death; patients with bulbar poliomyelitis, who may have apnea of brainstem origin, are clearly not dead. Although irreversible apnea has no strictly philosophical dimension, it is useful to include it in any concept of death. This is because of its obvious relation to cardiac function – if spontaneous breathing is lost the heart cannot long continue to function – and perhaps because of its cultural associations with the "breath of life". These aspects are addressed in the later discussion of how death has been envisaged in various cultures.

From as far back as medical records have been kept, it has been known that patients with severe head injuries or massive intracranial haemorrhage often die as a result of apnea: breathing stops before the heart does. In such cases, the pressure in the main (supratentorial) compartment of the skull becomes so great that brain tissue herniates through the tentorial opening, a bony and fibrous ring in the membrane that separates the spaces containing the cerebral hemispheres and the cerebellum. The brainstem runs through this opening, and a pressure cone formed by the herniated brain tissue may dislocate the brainstem downward and cause irreversible damage by squeezing it from each side. An early manifestation of such an event is a disturbance of consciousness; a late feature is permanent apnea. This was previously nature's way out.

With the widespread development of intensive care facilities in the 1950s and 1960s, more and more such moribund patients were rushed to specialized units and put on ventilators just before spontaneous breathing ceased. In some cases the effect was dramatic. When a blood clot could be evacuated, the primary brain damage and the pressure cone it had caused might prove reversible. Spontaneous breathing would return. In many cases, however, the massive, structural intracranial pathology was irremediable. The ventilator, which had taken over the functions of the paralysed respiratory centre, enabled oxygenated blood to be delivered to the heart, which went on beating. Physicians were caught up in a therapeutic dilemma partly of their own making: the heart was pumping blood to a dead brain. Sometimes the intracranial pressure was so high that the blood could not even enter the head. Modern technology was exacting a very high price: the beating-heart cadaver.

Brainstem death may also arise as an intracranial consequence of extracranial events. The main cause in such cases is circulatory arrest. The usual context is delayed or inadequate cardiopulmonary resuscitation following a heart attack. The intracranial repercussions depend on the duration and severity of impaired blood flow to the head. In the 1930s the British physiologist John Scott Haldane had emphasized that oxygen deprivation "not only stopped the machine, but wrecked the machinery". Circulatory arrest lasting two or three minutes can cause widespread and irreversible damage to the cerebral hemispheres while sparing the brainstem, which is more resistant to anoxia. Such patients remain in a "persistent vegetative state". They breathe and swallow spontaneously, grimace in response to pain, and are clinically and electrophysiologically awake, but they show no behavioural evidence of awareness. Their eyes are episodically open (so that the term coma is inappropriate to describe them), but their retained capacity for consciousness is not endowed with any content. Some patients have remained like this for many years. Such patients are not dead, and their prognosis depends in large part on the quality of the care they receive. The discussion of their management occasionally abuts onto controversies about euthanasia and the "right to die". These issues are quite different from that of the "determination of death", and failure to distinguish these matters has been the source of great confusion.

If circulatory arrest lasts for more than a few minutes, the brainstem – including its respiratory centre – will be as severely damaged as the cerebral hemispheres. Both the capacity for consciousness and the capacity to breathe will be irreversibly lost. The individual will then show all the clinical features of a dead brain, even if the heart can be restarted.

The diagnosis is not technically difficult. In more and more countries, it is made on purely clinical grounds. The aim of the clinical tests is not to probe every neuron within the intracranial cavity to see if it is dead – an impossible task – but to establish irreversible loss of brainstem function. This is the necessary and sufficient condition for irreversible unconsciousness and irreversible apnea, which together spell a dead patient. Experience has shown that instrumental procedures (such as electroencephalography and studies of cerebral blood flow) that seek to establish widespread loss of cortical function contribute nothing of relevance concerning the cardiac prognosis. Such tests yield answers of dubious reliability to what are widely felt to be the wrong questions. As the concept of brainstem death is relatively new, most countries rightly insist that the relevant examinations be carried out by physicians of appropriate seniority. These doctors (usually neurologists, anaesthetists, or specialists in intensive care) must be entirely separate from any who might be involved in using the patient's organs for subsequent transplants.

The diagnosis of brainstem death involves three stages. First, the cause of the coma must be ascertained, and it must be established that the patient (who will always have been in apneic coma and on a ventilator for several hours) is suffering from irremediable, structural brain damage. Damage is judged "irremediable" based on its context, the passage of time, and the failure of all attempts to remedy it. Second, all possible causes of reversible brainstem dysfunction, such as hypothermia, drug intoxication, or severe metabolic upset, must be excluded. Finally, the absence of all brainstem reflexes must be demonstrated, and the fact that the patient cannot breathe, however strong the stimulus, must be confirmed.

It may take up to 48 hours to establish that the preconditions and exclusions have been met; the testing of brainstem function takes less than half an hour. When testing the brainstem reflexes, doctors check for the following normal responses: (1) constriction of the pupils in response to light, (2) blinking in response to stimulation of the cornea, (3) grimacing in response to firm pressure applied just above the eye socket, (4) movements of the eyes in response to the ears being flushed with ice water, and (5) coughing or gagging in response to a suction catheter being passed down the airway. All responses have to be absent on at least two occasions. Apnea, which also must be confirmed twice, is assessed by disconnecting the patient from the ventilator. (Prior to this test, the patient is fully oxygenated by being made to breathe 100 per cent oxygen for several minutes, and diffusion oxygenation into the trachea is maintained throughout the procedure. These precautions ensure that the patient will not suffer serious oxygen deprivation while disconnected from the ventilator.) The purpose of this test is to establish the total absence of any inspiratory effort as the carbon dioxide concentration in the blood (the normal stimulus to breathing) reaches levels more than sufficient to drive any respiratory centre cells that may still be alive.

The patient thus passes through a tight double filter of preconditions and exclusions before they are even tested for the presence of a dead brainstem. This emphasis on strict preconditions and exclusions has been a major contribution to the subject of brainstem death, and it has obviated the need for ancillary investigations. Thousands of patients who have met criteria of this kind have had ventilation maintained: all have developed asystole within a few hours or a few days, and none has ever regained consciousness. There have been no exceptions. The relevant tests for brainstem death are carried out systematically and without haste. There is no pressure from the transplant team.

The developments in the idea and diagnosis of brainstem death came as a response to a conceptual challenge. Intensive-care technology had saved many lives, but it had also created many brain-dead patients. To grasp the implications of this situation, society in general – and the medical profession in particular – was forced to rethink accepted notions about death itself. The emphasis had to shift from the most common mechanism of death (i.e. irreversible cessation of the circulation) to the results that ensued when that mechanism came into operation: irreversible loss of the capacity for consciousness, combined with irreversible apnea. These results, which can also be produced by primary intracranial catastrophes, provide philosophically sound, ethically acceptable, and clinically applicable secular equivalents to the concepts of "departure of the soul" and "loss of the 'breath of life' ", which were important to earlier cultures, and for some, still remain so today.

14

ON THE CUTTING EDGE
OF BRAIN RESEARCH

Current Topics in Brain Research

Scientists in a broad range of fields, including cognitive neurobiology, evolutionary psychology, and psychopharmacology, are carrying out cutting-edge brain research that has the potential to save lives, to improve quality of life, and to enhance understanding of human information processing. The neuroscience of morality, deception, memory, consciousness, and emotion all have wide-ranging applications not only for the treatment of individuals suffering from mental disorders but also for the "reading" of mental processes and the anticipation of behaviours.

A subject of cognitive neurobiology and evolutionary psychology research with broad social and philosophical implications is that of morality. In fact, this research already shows signs that humans may indeed have a "universal moral grammar". Central to the neuroscientific study of morality is the repeated observation of affect-driven moral intuition. Research has shown that patients with damage to certain areas of the prefrontal cortex retain information-processing capabilities, such as IQ and explicit knowledge of right and wrong, but they show startling emotional deficits that impair their judgement and decision-making capabilities. Neuroimaging research conducted on healthy patients engaged in moral reasoning tasks has demonstrated that areas within the medial prefrontal cortex, such as the ventromedial prefrontal cortex and the medial frontal gyrus, are involved in moral reasoning. These areas integrate affect into judgement and decisions. Other imaging studies have demonstrated that affect originates in the amygdala and the frontal insula, which may function to sound "emotional alarms" that push judgement in one direction or another. When subjects are presented with difficult moral tasks that require them to do something normally considered repugnant in order to serve a greater good, they take longer to respond and show increased activity in their anterior cingulate cortex, an area associated with conflict. Subjects who found themselves able to override their intuitive "emotional alarms" exhibited increased activity in their dorsolateral prefrontal cortex. The analysis of these neuroscientific studies in conjunction with social psychological research has led to the theory that humans have a universal moral grammar that consists of five concepts: harm prevention, fairness, group loyalty, obedience to authority, and purity. There are also many indications that much of human moral reasoning is directly related to social interaction; however, this is difficult to demonstrate in controlled experiments because so much moral reasoning takes place in everyday life.

One avenue of research that has particular potential for practical application is "brain reading". Academic, military, and commercial scientists have been exploring the ability of functional magnetic resonance imaging (fMRI) scans and event-related potential (ERP) wave readings to detect memories, intentions, and deceptions. However, there are methodological and imaging difficulties associated with such research. In research on deception, one of the methodological difficulties is defining the parameters of the deception that will be studied. For instance, even though both are falsehoods, a lie is different from a delusion, because the speaker believes the delusion to be true. In addition, lying tends to be spontaneous and of the liar's own volition, whereas many experiments involving lying require the subject to lie as part of following the directions of the experimenter. These kinds of experiments have come under criticism for the very reason that they do not detect the essential character of real-world deception but instead detect some of the cognitive operations that factor into it. One of the cognitive operations that has been detected and localized in the brain is the estimation of risks and rewards, which has been localized to the medial frontal cortex and the anterior cingulate cortex. The anterior cingulate cortex is also associated with problem solving and motivation, as well as with the anticipation of consequences and the resolution of conflicting responses.

Other studies have demonstrated that the difference between rehearsed lies and spontaneous lies is, in fact, activation of the anterior cingulate cortex. This probably indicates that the liar is inhibiting the truth and checking to make sure that their created lie is different from the truth. A possible problem with this is that a subject could mimic the brain activity of a liar by thinking of some other conflict. If a truth-teller did this inadvertently they could be mistaken for a liar, and if a liar did this deliberately during a control condition, the liar would make their subsequent lies appear indistinguishable from the baseline. Another cognitive operation involved in deception is the mentalizing of the intentions of others in order to gain their trust. This operation is localized in the dorsomedial prefrontal cortex and the ventromedial prefrontal cortex. The dorsolateral prefrontal cortex has also been observed to be instrumental in the cognitive operation of response inhibition. In spite of these interesting discoveries, researchers need to overcome multiple methodological hurdles, including specificity of experimental conditions, extrapolation from large samples to individuals, and multiple interpretations of brain activation. If these hurdles can be overcome, research in deception could be used by police or for surveillance purposes.

If the application of brain imaging for the purposes of lie detection or memory detection is to remain a realistic goal, then serious strides will still need to be made in the understanding of how the brain processes information. Fascinating progress has already been made in the discovery of memory processing systems. An analysis of multiple studies of patients suffering from Alzheimer's disease, stroke, encephalitis (caused by herpes simplex virus), and semantic dementia (SD) have yielded an in-depth understanding of how the brain stores semantic memory and connects semantic memory to episodic memory. Semantic memories are general memories of objects, words, facts, and people that lack detail of time or place. In contrast, episodic memories are specific in their time and place. By comparing patients suffering from SD with patients suffering from the other three pathologies, researchers were able to find evidence supporting a "distributed-plus-hub" view of memory. The patients with SD had focal brain lesions in their anterior temporal lobes (ATL), whereas the other patients had either less severe or more widely distributed damage in their ATLs. Patients with SD also had very severe and very specific symptoms; for example, these patients demonstrated progressive deterioration of their vocabularies and knowledge of the properties of everyday objects, but their memories of recent events and other cognitive capacities remain perfectly intact. This has led some scientists to conclude that the ATL is a hub for amodal semantic memory, meaning that general conceptual information is not connected to one specific sensory feature of categories of objects. The ATL is adjacent to the anterior parts of the medial temporal lobe memory system, which processes episodic memory. The ATL is also adjacent to the amygdala and limbic structures, which process emotion and reward. By connecting these two regions, it is thought that the ATL acts as an amodal hub that draws general semantic information from episodic memories based on emotional salience. This is how the brain is able to categorize both oranges and bananas as "fruits", even though these objects differ in size, shape, and colour.

One of the more controversial avenues of neuroscientific research is the use of fMRI to detect the brain activity of individuals in a vegetative or minimally conscious state. A group of scientists used fMRI to study a woman who had sustained severe traumatic brain injury and had been unresponsive for five months. The researchers presented her with spoken sentences and with noise sequences. During the spoken sentences, she showed brain activation in the middle and superior temporal gyri that was equivalent to that of healthy volunteers presented with the same stimuli. When she was presented with sentences that contained ambiguous words such as *creak* (creek) she showed additional activity in the left inferior frontal region similar to that of healthy volunteers. She was also asked to imagine herself playing tennis and walking through all the rooms in her house. In the former, she showed activity in her supplementary motor area, and in the latter, she showed activity in the parahippocampal gyrus, the posterior parietal cortex, and the lateral premotor cortex. These patterns of activation were indistinguishable from those of healthy volunteers. However, it remains controversial whether brain activity means that she is in fact conscious. The scientists performing the research claim that the woman is conscious because she followed their directions, which would mean that she is acting out conscious intentions. Others argue that she is not conscious, because she did not report a mental state and did not *spontaneously* act out any conscious intentions. In addition, many patients in vegetative states show local and specific patterns of brain activation but do not show the long-range neural integration that is a hallmark of consciousness. The ethical, philosophical, and scientific debates on this particular issue will most likely continue, even as more scientific information becomes available.

Treatments for emotional disorders, such as depression, anxiety, OCD, and PTSD, have undergone tremendous progress in recent years. Perhaps what is most fascinating about these treatments is that they are not just pharmacological in nature. For example, electroconvulsive therapy (ECT), transcranial magnetic stimulation (TMS), and deep brain stimulation (DBS) all show tremendous potential for alleviating multiple emotional disorders. ECT, although its mechanisms are not completely understood, has proven to be one of the most effective treatments for depression. TMS, which consists of placing a large magnet over the scalp and inducing short electromagnetic pulses, has also shown promise. However, the specific treatment parameters and patient characteristics needed for optimal benefit from TMS still need to be investigated. DBS, which entails implanting an electrode directly into the brain, has been successful in treating Parkinson's disease, and there is some evidence that DBS alleviates depressive symptoms when implanted in the caudate nucleus, similar to Parkinson treatments. However, DBS is much more effective in alleviating depression when it is implanted in prefrontal cortex area Cg25. One study even demonstrated that DBS also decreases OCD symptoms when implanted in the internal capsule extending into the ventral capsule/ventral striatum (VC/VS). The drug D-cycloserine (DCS) significantly reduces fear and anxiety in patients suffering from phobias or social anxiety disorder when it is used in conjunction with exposure therapy, which gradually exposes people to their fears. In addition, propranolol, a beta-blocker used for hypertension, has shown promise in small studies and is currently in clinical trials for treatment of PTSD. Propranolol works by preventing the consolidation of the emotional aspects of traumatic memories when administered after a trauma.

Finally, two areas of study that show exceptional potential and also revolutionize the understanding of how the brain works are the areas of neuroplasticity and stem cell therapy. Up until the 1970s, there was consensus in the neuroscientific community that aside from learning and memory, the brain never changed physically throughout the course of life, and, once damaged, the brain was unable to recover. However, research into stroke recovery and phantom limb syndrome has shown that the brain can adapt its functioning and architecture depending on damage, behaviour, or other environmental conditions – an ability dubbed neuroplasticity. Furthermore, in the late 1990s, scientists discovered neural stem cells – cells that are found in the brain and are actually capable of generating new neurons and glial cells as people age. By extracting, cultivating, and transplanting these cells, or by inducing their propagation within the body, scientists hope to keep the brain forever young and intact. However, there remain many scientific hurdles and ethical controversies between present-day stem cell research and the miracle therapies of the future.

What is Neuroplasticity?

Neuroplasticity is the capacity of local neuronal networks and neural systems to change their topography and local architecture in response to new information, sensory stimulation, development, damage, or dysfunction. Although neural networks also exhibit modularity and carry out specific functions, they retain the capacity to deviate from their usual functions and to reorganize themselves. In fact, for many years, it was considered dogma in the neurosciences that certain functions were hard-wired in specific, localized regions of the brain and that any incidents of brain change or recovery were mere exceptions to the rule. However, since the 1970s and 1980s, neuroplasticity has gained wide acceptance throughout the scientific community as a complex, multifaceted, fundamental property of the brain.

Rapid change or reorganization of the brain's cellular or neural networks can take place in many different forms and under many different circumstances. Developmental plasticity occurs when neurons in the young brain rapidly sprout branches and form synapses. Then, as the brain begins to process sensory information, some of these synapses strengthen and others weaken. Eventually, some unused synapses are eliminated completely, a process known as synaptic pruning, which leaves behind efficient networks of neural connections. Other forms of neuroplasticity operate by much the same mechanism but under different circumstances and sometimes only to a limited extent. These circumstances include changes in the body, such as the loss of a limb or sense organ, which subsequently alter the balance of sensory activity received by the brain. Another circumstance is when sensory information is reinforced through experience, such as in learning and memory. The final major circumstance is when the brain experiences actual physical damage from a trauma such as a stroke and attempts to compensate for lost activity. Today, it has become apparent that the same brain mechanisms – adjustments in the strength or number of synapses between neurons – operate in all these situations. Sometimes this happens naturally, which can result in positive or negative reorganization, but other times, behavioural techniques or brain–machine interfaces can be used to harness the power of neuroplasticity for therapeutic purposes. In some cases such as stroke recovery, natural adult neurogenesis can also play a role. As a result, neurogenesis has spurred an interest in stem cell research, which could lead to an enhancement of neuro-genesis in adults who suffer from stroke, Alzheimer's disease, Parkinson's disease, or depression.

Types of Cortical Neuroplasticity

Developmental plasticity occurs most profoundly in the first few years of life as neurons grow very rapidly and send out multiple branches, ultimately forming too many connections. In fact, at birth, each neuron in the cerebral cortex has about 2,500 synapses. By the time an infant is two or three years old, the number of synapses is approximately 15,000 per neuron. This amount is about twice that of the average adult brain. The connections that are not reinforced by sensory stimulation eventually weaken, and the connections that are reinforced become stronger. Eventually, efficient pathways of neural connections are carved out. Throughout the life of a human or other mammal, these neural connections are fine-tuned through the organism's interaction with its surroundings. In addition, a massive "pruning back" of excess neurons occurs during adolescence. However, early childhood is known as a critical period, during which the nervous system must receive certain sensory inputs in order to develop properly. Once a critical period such as early childhood ends, there is a precipitous drop in the number of connections that are maintained, and the ones that do remain are the ones that have been strengthened by the appropriate sensory experiences.

American neuroscientist Jordan Grafman has identified four other types of neuroplasticity, known as homologous area adaptation, compensatory masquerade, cross-modal reassignment, and map expansion. Homologous area adaptation occurs during the critical period of early development. When a particular brain module becomes damaged in early life, its normal operations have the ability to shift to brain areas that do not include the affected module. The function is often shifted to a module in the matching, or homologous, area of the opposite brain hemisphere. The downside to this form of neuroplasticity is that it may come at costs to functions that are normally stored in the module but now have to make room for the new functions. An example of this is when the right parietal lobe becomes damaged early in life, and the left parietal lobe takes over visuospatial functions at the cost of impaired arithmetical functions, which the left parietal lobe usually carries out exclusively. Timing is also a factor in this process, because a child learns how to navigate physical space before they learn arithmetic.

The second type of neuroplasticity, compensatory masquerade, can simply be described as the brain figuring out an alternative strategy for carrying out a task when the initial strategy cannot be followed due to impairment. One example is when someone attempts to navigate from one location to another. Most people, to a greater or lesser extent, have an intuitive sense of direction and distance that they employ for navigation. However, if a person suffers some form of brain trauma that impairs spatial sense, they resort to another strategy for spatial navigation such as memorizing landmarks. The only change that occurs in the brain is a reorganization of pre-existing neuronal networks.

The third form of neuroplasticity, cross-modal reassignment, entails the introduction of new inputs into a brain area deprived of its main inputs. A classic example of this is the ability of adults who have been blind since birth to have touch, or somatosensory, input redirected to their visual cortices in the occipital lobes of their brains, specifically, in an area known as V1. Sighted people, however, do not display any V1 activity when presented with similar touch-oriented experiments. This occurs because neurons communicate with one another in the same abstract "language" of electrochemical impulses regardless of sensory modality. Moreover, all the sensory cortices of the brain – visual, auditory, olfactory (smell), gustatory (taste), and somatosensory (touch) – have a similar six-layer processing structure. Because of this, the visual cortices of blind people can still carry out the cognitive functions of creating representations of the physical world but base these representations on input from another sense, namely, touch. This is not, however, simply an instance of one area of the brain compensating for a lack of vision; it is a change in the actual functional assignment of a local brain region. There are limits to this type of neuroplasticity; some cells such as the colour-processing cells of the visual cortex are so specialized for visual input that there is very little likelihood that they would accept input from other senses.

Map expansion, the fourth type of neuroplasticity, entails the flexibility of local brain regions that are dedicated to performing one type of function or storing a particular form of information. The arrangement of these local regions in the cortex is referred to as a "map". When one function is carried out frequently enough through repeated behaviour or stimulus, the region of the cortical map dedicated to this function grows and shrinks as an individual "exercises" this function. This phenomenon usually takes place during the learning and practising of a skill such as playing a musical instrument. Specifically, the region grows as the individual gains implicit familiarity with the skill and then shrinks to baseline once the learning becomes explicit. But as one continues to develop the skill over repeated practice, the region retains the initial enlargement.

Map expansion neuroplasticity also underlies the phenomenon of phantom limb syndrome. This phenomenon was first proposed by American physician Silas Weir Mitchell, who tended to wounded soldiers in Philadelphia during the American Civil War. In the 1990s, American neuroscientist Vilayanur S. Ramachandran applied theories of neuroplasticity to try to explain this strange syndrome, which is characterized by the ability of amputees to still feel sensations and even pain in limbs that no longer exist. Ramachandran discovered that when he touched the side of an amputee's face with a cotton swab, the amputee also felt a sensation in their phantom arm. After scanning an amputee's brain using magnetoencephalography (MEG), Ramachandran also discovered that the amputee's face brain map had taken over the adjacent area of the arm and hand brain maps. Ramachandran eventually concluded that after the arm was amputated, the arm and hand brain maps were starved for input and sent out growth factors that attracted neuronal sprouting from the face map stored in an adjacent area of the brain.

Brain–Machine and Brain–Computer Interfaces

Some of the earliest applied research in neuroplasticity was carried out in the 1960s, when scientists attempted to develop machines that interface with the brain in order to help blind people. In 1969, American neurobiologist Paul Bach-y-Rita published a short article detailing the workings of such a machine. The machine consisted of a metal plate with 400 vibrating stimulators. The plate was attached to the back of a chair so that the sensors could touch the skin of the patient's back. A camera was placed in front of the patient and was connected to the vibrators. The camera acquired images of the room and translated them into patterns of vibration, which represented the physical space of the room and the objects in it. After patients gained some familiarity with the device, their brains were able to construct mental representations of physical spaces and physical objects. Thus, instead of visible light stimulating their retinas and creating a mental representation of the world, vibrating stimulators triggered the skin of their backs to create a representation in their visual cortices. A similar device exists today, only the camera fits inside a pair of glasses, and the sensory surface fits on the tongue. The brain can do this because it "speaks" in the same neural "language" of electrochemical signals regardless of what kinds of environmental stimuli are interacting with the body's sense organs.

Neuroscientists have recently started to develop machines that bypass external sense organs and actually interface directly with the brain. For example, researchers implanted a device that monitored neuronal activity in the brain of a female macaque monkey. The monkey used a joystick to move a cursor around a screen, and a computer monitored and compared the movement of the cursor with the activity in the monkey's brain. Once the computer had effectively correlated the monkey's brain signals for speed and direction to the actual movement of the cursor, the computer was able to translate these movement signals from the monkey's brain to the movement of a robot arm in another room. Thus, the monkey became capable of moving a robot arm with its thoughts. However, the major finding of this experiment was that as the monkey learned to move the cursor with its thoughts, the signals in the monkey's motor cortex became less representative of the movements of the monkey's actual limbs and more representative of the movements of the cursor. This means that the motor cortex does not control the details of limb movement directly but instead controls the abstract parameters of movement, regardless of the connected apparatus that is actually moving. This has also been observed in humans whose motor cortices can easily be manipulated into incorporating a tool or prosthetic limb into the brain's body image through both somatosensory and visual stimuli.

For humans, however, less invasive forms of brain–computer interfaces are more conducive to clinical application. For example, researchers have demonstrated that real-time visual feedback from fMRI can enable patients to retrain their brains and therefore improve brain function. Patients with emotional disorders have been trained to self-regulate a region of the brain known as the amygdala by self-inducing sadness and monitoring the activity of the amygdala on a real-time fMRI readout. Stroke victims have also been able to reacquire lost functions through self-induced mental practice and mental imagery. This kind of therapy takes advantage of neuroplasticity in order to reactivate damaged areas of the brain or deactivate overactive areas of the brain. Today, researchers are investigating the efficacy of these forms of therapy for individuals affected by stroke, emotional disorders, chronic pain, psychopathy, or social phobia.

Neural Stem Cells

For years it was thought that the brain was a closed fixed system. Even the renowned Spanish neuroanatomist Santiago

Ramón y Cajal, who won the Nobel Prize in 1906 for establishing the neuron as the fundamental cell of the brain, was unaware of neurogenesis during his otherwise remarkable career. There were only a handful of discoveries, primarily in rats, birds, and primates, in the latter half of the twentieth century that hinted at the regenerative capability of brain cells. During this time, scientists assumed that once the brain was damaged or began to deteriorate it could not regenerate new cells in the way that other types of cells, such as liver and skin cells, are able to regenerate. The generation of new brain cells in the adult brain was thought to be impossible because a new cell could never fully integrate itself into a complex system like the brain. It was not until 1998 that neural stem cells (NSCs) were discovered in humans.

NSCs were discovered in a region of the brain called the hippocampus, a region known to be instrumental in the formation of memories. Eventually, NSCs were also found to be active in the olfactory bulbs (an area that processes smell) and dormant and inactive in the septum (an area that processes emotion), the striatum (an area that processes movement), and the spinal cord. The reason that these stem cells are referred to as "neural" is because they have the potential to develop into either neurons or glial cells, which insulate neurons and enhance the speed at which they send signals. Today, scientists are doing research on pharmaceuticals that could activate dormant stem cells in case the areas where they are located become damaged. Other avenues of research seek to figure out ways to transplant stem cells into damaged areas and to coax them to migrate to damaged areas. Still other stem cell researchers seek to take stem cells from other sources (i.e. embryos) and to influence these cells to develop into neurons or glial cells. The most controversial of these stem cells are the ones that are procured from human embryos. These stem cells are controversial because they require the destruction of human embryos in order to be obtained. Scientists have been able to create induced pluripotent stem cells by reprogramming adult somatic cells (cells of the body, excluding sperm and egg cells) through the introduction of certain regulatory genes. However, the generation of these cells requires the use of a retrovirus, and therefore these cells have the potential to introduce harmful cancer-causing viruses into patients. While further research is needed to develop better methods of isolating and generating embryonic stem cells, these cells possess amazing potential because they are capable of being turned into any type of cell found in the human body.

Stroke recovery is one area of research where much has been discovered about the promise and the complexities of stem cell therapy. For example, two main approaches can be taken to stem cell therapy. One is the endogenous approach, and the other is the exogenous approach. The endogenous approach relies on stimulating adult NSCs within the patient's own body. These stem cells are found in two zones of the dentate gyrus in the brain, as well as in the striatum, the neocortex, and the spinal cord. In rat models, growth factors, such as fibroblast growth factor-2, vascular endothelial growth factor (VEGF), brain-derived neurotrophic factor (BDNF), and erythropoietin, have been administered after strokes in an effort to induce or to enhance neurogenesis, thereby staving off brain damage and spurring functional recovery. The most promising growth factor in the rat models was erythropoietin, which promotes neural progenitor cell proliferation and has been shown to induce neurogenesis and functional improvement following embolic stroke in rats. This was followed by clinical trials in which erythropoietin was administered to a small sample of stroke patients, who eventually showed dramatic improvements over individuals in the placebo group. Erythropoietin has also shown promise in schizophrenics and in patients with multiple sclerosis. However, further studies need to be performed in larger groups of patients in order to confirm the efficacy of erythropoietin.

Exogenous stem cell therapies rely upon the extraction, *in vitro* cultivation, and subsequent transplantation of stem cells into the regions of the brain affected by stroke. Studies have shown that adult neural stem cells can be obtained from the dentate gyrus, hippocampus, cortex, and subcortical white matter. However, actual transplantation studies have only been carried out in rats with spinal cord injury using stem cells that had been biopsied from the subventricular zone of the adult brain. Fortunately, there were no functional deficits as a result of the biopsy. There have also been studies in rats in which embryonic stem cells or fetal-derived neural stem and progenitor cells (undifferentiated cells; similar to stem cells but with narrower differentiation capabilities) have been transplanted into regions of the brain damaged by stroke. In these studies, the grafted neural stem cells successfully differentiated into neurons and glial cells, and there was some functional recovery. The major caveat, however, with exogenous therapies is that scientists have yet to fully understand the underlying mechanisms of differentiation of the progenitor cells and their integration into existing neuronal networks. Scientists and clinicians have yet to understand how to control the proliferation, migration, differentiation, and survival of neural stem cells (NSCs) and their progeny. This is due to the fact that NSCs are partially regulated by the specialized microenvironment – or "niche" – in which these cells reside.

There has also been research into hematopoietic stem cells (HSCs), which usually differentiate into blood cells but can also be transdifferentiated into neural lineages. These HSCs can be found in bone marrow, umbilical cord blood, and peripheral blood cells. Interestingly, these cells have been found to be spontaneously mobilized by certain types of strokes and can also be further mobilized by granulocyte colony stimulating factor (G-CSF). Studies of G-CSF in rats have shown that G-CSF can lead to functional improvement following stroke, and clinical trials in humans look promising. Exogenous studies have also been carried out in rats with HSCs. The HSCs were administered locally at the site of damage in some studies or administered systemically through intravenous transplantation in other studies. The latter procedure is simply more feasible, and the most effective HSCs seem to be HSCs derived from the peripheral blood.

The research that has been done on stem cell therapies for epilepsy and Parkinson's disease also demonstrate the promise and difficulty of properly cultivating stem cells and introducing them into a living system. With regard to embryonic stem cells (ESCs), studies have shown that they are capable of being differentiated into dopaminergic neurons, spinal motor neurons, and oligodendrocytes. In studies aimed at treating epilepsy, mouse embryonic stem cell-derived neural precursors (ESNs) were transplanted into the hippocampi of chronically epileptic rats and control rats. After transplantation, no differences where found in the functional properties of the ESNs as they all displayed the synaptic properties characteristic of neurons. However, it still remains to be seen whether ESNs have the ability to survive for prolonged periods in the epileptic hippocampus, to differentiate into neurons with the proper hippocampal functions, and to suppress learning and memory deficits during chronic epilepsy. NSCs, on the other hand, have already been observed to survive and to differentiate into different functional forms of neurons in rats. However, it remains unclear whether they can differentiate into the different functional forms in appropriate amounts and whether they can synapse properly with hyperexcitable neurons in order to inhibit them to curb seizures. These details will need to be studied thoroughly in animal models before clinical trials can be conducted in humans.

Treatments for Parkinson's disease also face similar promise and obstacles. Clinical research has been carried out on the transplantation of human fetal mesencephalic tissue into the striata of Parkinson patients. However, this tissue is of limited availability, which is what makes embryonic stem cell transplantation more appealing. Indeed, research has already shown that transplantable dopaminergic neurons – the kind of neurons affected in Parkinson's disease – can be generated from mouse, primate, and human ESCs. The one major difference between mouse and human ESCs, however, is that human ESCs take much longer to differentiate (up to 50 days). Also, differentiation programmes for human ESCs require the introduction of animal serum in order to propagate, which might violate certain medical regulations, depending on the country. Researchers will also need to figure out a way to get ESC-derived dopaminergic progenitor cells to survive for a longer period of time after transplantation. Finally, there is the issue of the purity of ESC-derived cell populations – all the cells must be certified as dopaminergic precursor cells before they can be safely transplanted. Nevertheless, differentiation and purification techniques are improving with each study. Indeed, generation of large banks of pure and specific cell populations for human transplantation remains an attainable goal.

GLOSSARY

A

acetylcholine A neurotransmitter active both in the brain, where it regulates memory, and in the peripheral nervous system, where it controls the actions of skeletal and smooth muscle.

action potential An electrical charge travels along the axon to the neuron's terminal, where it triggers the release of a neurotransmitter. This occurs when a neuron is activated and temporarily reverses the electrical state of its interior membrane from negative to positive.

adrenal cortex An endocrine organ that secretes corticosteroids for metabolic functions; for example, in response to stress.

adrenal medulla An endocrine organ that secretes epinephrine and norepinephrine in concert with the activation of the sympathetic nervous system; for example, in response to stress.

agonist A neurotransmitter, drug, or other molecule that stimulates receptors to produce a desired reaction.

Alzheimer's disease The major cause of dementia most prevalent in the elderly. The disease is characterized by death of neurons in the hippocampus, cerebral cortex, and other brain regions.

amino acid transmitters The most prevalent neurotransmitters in the brain, these include glutamate and aspartate, which have excitatory actions on nerve cells, and glycine and gamma-amino butyric acid (GABA), which also have inhibitory actions on nerve cells.

amnesia Defined as loss of memory occurring most often as a result of damage to the brain from trauma, stroke, Alzheimer's disease, alcohol and drug toxicity, or infection.

amygdala A structure in the forebrain that is an important component of the limbic system and plays a central role in emotional learning, particularly within the context of fear.

androgens Sex steroid hormones, including testosterone, found in higher levels in males than females. They are responsible for male sexual maturation.

antagonist A drug or other molecule that blocks receptors. Antagonists inhibit the effects of agonists.

aphasia Disturbance in language comprehension or production, often as a result of a stroke.

apoptosis Programmed cell death induced by specialized biochemical pathways, often serving a specific purpose in the development of the animal.

artificial intelligence (AI) The ability of a digital computer or computer-controlled robot to perform tasks commonly associated with intelligent beings. The term is frequently applied to the project of developing systems endowed with the intellectual processes characteristic of humans, such as the ability to reason, discover meaning, generalize, or learn from past experience.

astrocytes A sub-type of glial cell that have many functions, such as aiding the blood–brain barrier and healing scarring in the brain.

auditory nerve A bundle of nerve fibres extending from the cochlea of the ear to the brain that contains two branches: the cochlear nerve, which transmits sound information, and the vestibular nerve, which relays information related to balance.

autonomic nervous system A part of the peripheral nervous system responsible for regulating the activity of internal organs. It includes the sympathetic and parasympathetic nervous systems.

axon The fibre-like extension of a neuron by which it sends information to target cells.

axon hillock The region where an axon rises from the soma in a neuron.

B

basal ganglia Clusters of neurons, which include the caudate nucleus, putamen, globus pallidus, and substantia nigra, located deep in the brain. Play an important role in the initiation of movements. Cell death in the substantia nigra contributes to Parkinson's disease.

brainstem The major route by which the forebrain sends information to and receives information from the spinal cord and peripheral nerves. The brainstem controls, among other things, respiration and the regulation of heart rhythms. It comprises the midbrain, pons, and medulla oblongata.

Broca's area The brain region located in the frontal lobe of the left hemisphere that is important for the production of speech.

C

catecholamines The neurotransmitters dopamine, epinephrine, and norepinephrine, which are active in both the brain and the peripheral sympathetic nervous system. These three molecules have certain structural similarities and are part of a larger class of neurotransmitters known as monoamines.

central nervous system The brain and the spinal cord make up the central nervous system.

central sulcus Another name for the fissure of Ronaldo. The central sulcus and the lateral sulcus are two major furrows that divide the cerebral hemisphere of the brain into the four lobes: frontal, parietal, temporal, and occipital.

cerebellum A large structure located at the roof of the hindbrain that helps control the coordination of movement by making connections to the pons, medulla, spinal cord, and thalamus. It also may be involved in aspects of motor learning.

cerebral cortex The outermost layer of the cerebral hemispheres of the brain. It is largely responsible for all forms of conscious experience, including perception, emotion, thought, and planning.

cerebral hemispheres The two specialized halves of the brain. For example, in right-handed people, the left hemisphere is specialized for speech, writing, language, and calculation; the right hemisphere is specialized for spatial abilities, visual face recognition, and some aspects of music perception and production.

cerebral ventricles Four cavities in the brain that are continuous with the central canal of the spinal cord.

cerebrospinal fluid A liquid found within the ventricles of the brain and the central canal of the spinal cord.

cerebrum The evolutionarily newest and most highly developed part of the brain that is in charge of complex functions such as language. This is generally considered to be the structure that separates humans from animals.

cholecystokinin A hormone released from the lining of the stomach during the early stages of digestion that acts as a powerful suppressant of normal eating. It also is found in the brain.

circadian rhythm A cycle of behaviour or physiological change lasting approximately 24 hours.

classical conditioning Learning in which a stimulus that naturally produces a specific response (unconditioned stimulus) is repeatedly paired with a neutral stimulus (conditioned stimulus). As a result, the conditioned stimulus can come to evoke a response similar to that of the unconditioned stimulus.

cochlea A snail-shaped, fluid-filled organ of the inner ear responsible for transducing motion into neurotransmission to produce an auditory sensation.

cognition The process or processes by which an organism gains knowledge or becomes aware of events or objects in its environment and uses that knowledge for comprehension and problem solving.

cone A primary receptor cell for vision located in the retina. It is sensitive to colour and is used primarily for daytime vision.

corpus callosum The large bundle of nerve fibres linking the left and right cerebral hemispheres.

cortisol A hormone manufactured by the adrenal cortex. In humans, cortisol is secreted in the greatest quantities before dawn, readying the body for the activities of the coming day.

D

delusion A false or irrational belief that is firmly held despite obvious or objective evidence to the contrary.

dendrite A tree-like extension of the neuron cell body. The dendrite is the primary site for receiving and integrating information from other neurons.

depression A mental disorder characterized by depressed mood and abnormalities in sleep, appetite, and energy level.

diencephalon A region of the brain beneath the cerebrum, on top of the brainstem. It includes the epithalamus, thalamus, hypothalamus and subthalamus. Its various functions include acting as a relay system between incoming sensory input and other areas of the brain, and as a site for interaction between the central nervous and endocrine systems. It also has a role in the limbic system.

dopamine A catecholamine neurotransmitter known to have varied functions depending on where it acts. Dopamine-containing neurons in the substantia nigra of the brainstem project to the caudate nucleus and are destroyed in Parkinson's disease. Dopamine is thought to regulate key emotional responses such as reward and plays a role in schizophrenia and drug abuse.

dorsal horn An area of the spinal cord where many nerve fibres from peripheral pain receptors meet other ascending and descending nerve fibres.

drug addiction Loss of control over drug intake or compulsive seeking and taking of drugs, despite adverse consequences.

E

endocrine organ An organ that secretes a hormone directly into the bloodstream to regulate cellular activity of certain other organs.

endorphins Neurotransmitters produced in the brain that generate cellular and behavioural effects like those of morphine.

epilepsy A disorder characterized by repeated seizures, which are caused by abnormal excitation of large groups of neurons in various brain regions. Epilepsy can be treated with many types of anticonvulsant medications.

epinephrine A hormone released by the adrenal medulla and specialized sites in the brain, that acts with norepinephrine to affect the sympathetic division of the autonomic nervous system. Also known as adrenaline.

epithalamus The epithalamus is represented mainly by the pineal gland, which lies in the midline posterior and dorsal to the third ventricle. This gland synthesizes melatonin and enzymes sensitive to daylight.

evoked potentials A measure of the brain's electrical activity in response to sensory stimuli. This is obtained by placing electrodes on the surface of the scalp (or more rarely, inside the head), repeatedly administering a stimulus, and then using a computer to average the results.

excitation A change in the electrical state of a neuron that is associated with an enhanced probability of action potentials.

F

follicle-stimulating hormone A hormone released by the pituitary gland that stimulates the production of sperm in the male and growth of the follicle (which produces the egg) in the female.

forebrain The largest part of the brain, which includes the cerebral cortex, the diencephalon, and basal ganglia. The forebrain is credited with the highest intellectual functions.

frontal lobe One of the four divisions (the other lobes are the parietal, temporal, and occipital) of each hemisphere of the cerebral cortex. The frontal lobe has a role in controlling movement and in the planning and coordinating of behaviour.

G

gamma-amino butyric acid (GABA) An amino acid transmitter in the brain whose primary function is to inhibit the firing of nerve cells.

Gestalt In the language of Gestalt psychologists, immediate human experience is of organized wholes (*Gestalten*), not of collections of elements. A major goal of Gestalt theory in the twentieth century was to specify the brain processes that might account for the organization of perception. Gestalt theorists rejected the earlier assumption that perceptual organization was the product of learned relationships (associations), the constituent elements of which were called simple sensations.

glia Specialized cells that nourish and support neurons.

glutamate An amino acid neurotransmitter that acts to excite neurons. Glutamate stimulates N-methyl-D-aspartate (NMDA) and alpha-amino-3-hydroxy-5-methylisoxazole-4-propionic acid (AMPA). AMPA receptors have been implicated in activities ranging from learning and memory to development and specification of nerve contacts in developing animals. Stimulation of NMDA receptors may promote beneficial changes, whereas overstimulation may be a cause of nerve cell damage or death in neurological trauma and stroke.

gonad Primary sex gland: testis in the male and ovary in the female.

growth cone A distinctive structure at the growing end of most axons. It is the site where new material is added to the axon.

H

hindbrain Comprises the cerebellum, medulla oblongata, and the pons.

hippocampus A seahorse-shaped structure located within the brain and considered an important part of the limbic system. One of the most studied areas of the brain, it functions in learning, memory, and emotion.

hormones Chemical messengers secreted by endocrine glands to regulate the activity of target cells. They play a role in sexual development, calcium and bone metabolism, growth, and many other activities.

Huntington's disease A movement disorder caused by the death of neurons in the basal ganglia and other brain regions. It is characterized by abnormal movements called chorea – sudden, jerky movements without purpose.

hypersomnia Excessive sleep, sometimes bordering on coma, e.g. narcolepsy.

hyposomnia Too little sleep, sometimes preferred to "insomnia", or "lack of sleep", because some sleep invariably is present.

hypothalamus A complex brain structure composed of many nuclei with various functions, including regulating the activities of internal organs, monitoring information from the autonomic nervous system, controlling the pituitary gland, and regulating sleep and appetite.

I

inhibition A synaptic message that prevents a recipient neuron from firing.

interneuron Neurons that have two axons rather than an axon and a dendrite. While one of these axons extends toward the skin or muscle, the other extends toward the spinal cord. They send information between motor neurons and sensory neurons

ions Electrically charged atoms or molecules.

ion channels Proteins that form the pores in cell membranes and hence regulate the flow of ions across the membrane.

insomnia Lack of sleep. See also **hyposomnia**.

L

lateral sulcus Also called the fissure of Sylvius. The lateral sulcus and the central sulcus (or fissure of Ronaldo) are two major furrows that divide the cerebral hemisphere of the brain into the four lobes: frontal, parietal, temporal, and occipital.

limbic system A group of brain structures – including the amygdala, hippocampus, septum, basal ganglia, and others – that help regulate the expression of emotion and emotional memory.

long-term memory The final phase of memory, in which information storage may last from hours to a lifetime.

M

mania A mental disorder characterized by excessive excitement, exalted feelings, elevated mood, psychomotor overactivity, and overproduction of ideas. It may be associated with psychosis; for example, delusions of grandeur.

medulla oblongata Is the lowest part of the brain and the lowest portion of the brainstem.

melatonin Produced from serotonin, melatonin is released by the pineal gland into the bloodstream. Melatonin affects physiological changes related to time and lighting cycles.

memory consolidation The physical and psychological changes that take place as the brain organizes and restructures information to make it a permanent part of memory.

meninges the system of membranes that envelop the central nervous system.

metabolism The sum of all physical and chemical changes that take place within an organism and all energy transformations that occur within living cells.

microglial cells A small sub-type of glial cell that has a dark cytoplasm and a dark nucleus. Their main role is to act as an active immune defence in the central nervous system.

midbrain The most anterior segment of the brainstem. With the pons and medulla, the midbrain is involved in many functions, including regulation of heart rate, respiration, pain perception, and movement.

mitochondria Small cylindrical organelles inside cells that provide energy for the cell by converting sugar and oxygen into special energy molecules, called ATP.

monoamine oxidase (MAO) The brain and liver enzyme that normally breaks down the catecholamines norepinephrine, dopamine, and epinephrine, as well as other monoamines such as serotonin.

motor neuron A neuron that carries information from the central nervous system to muscle.

myasthenia gravis A disease in which acetylcholine receptors on muscle cells are destroyed so that muscles can no longer respond to the acetylcholine signal to contract. Symptoms include muscular weakness and progressively more common bouts of fatigue. The disease's cause is unknown but it is more common in females than in males; it usually strikes between the ages of 20 and 50.

myelin Compact fatty material that surrounds and insulates the axons of some neurons.

N

narcolepsy Irresistible brief episodes of sleep in the daytime.

necrosis Cell death due to external factors, such as lack of oxygen or physical damage, that disrupt the normal biochemical processes in the cell.

nerve growth factor A substance whose role is to guide neuronal growth during embryonic development, especially in the peripheral nervous system. Nerve growth factor also probably helps sustain neurons in the adult.

neuroglia *see* glia

neuromodulator Substances that do not actively activate ion-channel receptors but that, acting together with neurotransmitters, enhance the excitatory or inhibitory responses of the receptors.

neuron A nerve cell specialized for the transmission of information and characterized by long, fibrous projections called axons and shorter, branch-like projections called dendrites.

neuroplasticity A general term used to describe the adaptive changes in the structure or function of nerve cells or groups of nerve cells in response to injuries to the nervous system or alterations in patterns of their use and disuse.

neurotransmitter A chemical released by neurons at a synapse for the purpose of relaying information to other neurons via receptors.

nociceptors In animals, nerve endings that signal the sensation of pain. In humans, they are called pain receptors.

norepinephrine A catecholamine neurotransmitter, produced both in the brain and in the peripheral nervous system. Norepinephrine is involved in arousal and in regulation of sleep, mood, and blood pressure.

O

occipital lobe One of the four subdivisions of the cerebral cortex. The occipital lobe plays a role in processing visual information.

oestrogens A group of sex hormones found more abundantly in females than males. They are responsible for female sexual maturation and other functions.

oligodendrocytes A sub-type of glial cell that has a main function of exclusively insulating the axons found in the central nervous system.

organelles Small structures within a cell that maintain the cell and do the cell's work.

P

parasympathetic nervous system A branch of the autonomic nervous system concerned with the conservation of the body's energy and resources during relaxed states.

parietal lobe One of the four subdivisions of the cerebral cortex. The parietal lobe plays a role in sensory processes, attention, and language.

Parkinson's disease A movement disorder caused by death of dopamine neurons in the substantia nigra, located in the midbrain. Symptoms include tremor, shuffling gait, and general paucity of movement.

peptides Chains of amino acids that can function as neurotransmitters or hormones.

peripheral nervous system A division of the nervous system consisting of all nerves that are not part of the brain or spinal cord.

phosphorylation A process that modifies the properties of neurons by acting on an ion channel, neurotransmitter receptor, or other regulatory protein. During phosphorylation, a phosphate molecule is placed on a protein and results in the activation or inactivation of the protein. Phosphorylation is believed to be a necessary step in allowing some neurotransmitters to act and is often the result of second-messenger activity.

pineal gland An endocrine organ found in the brain. In some animals, the pineal gland serves as a light-influenced biological clock.

pituitary gland An endocrine organ closely linked with the hypothalamus. In humans, the pituitary gland is composed of two lobes and secretes several different hormones that regulate the activity of other endocrine organs throughout the body.

pons A part of the hindbrain that, with other brain structures, controls respiration and regulates heart rhythms. The pons is a major route by which the forebrain sends information to and receives information from the spinal cord and peripheral nervous system.

psycholinguistics (or psychology of language) Study of the psychological and neurobiological factors that enable humans to acquire, use, and understand language. Initial forays into psycholinguistics were largely philosophical ventures, due mainly to a lack of cohesive data on how the human brain functioned. Modern research makes use of biology, neuroscience, cognitive science, and information theory to study how the brain processes language. There are a number of subdisciplines; for example, as non-invasive techniques for studying the neurological workings of the brain become more and more widespread, neurolinguistics has become a field in its own right.

psychosis A severe symptom of mental disorders characterized by an inability to perceive reality. Psychosis can occur in many conditions, including schizophrenia, mania, depression, and drug-induced states.

psychotherapy (also called counselling) Any form of treatment for psychological, emotional, or behaviour disorders in which a trained person establishes a relationship with one or several patients for the purpose of modifying or removing existing symptoms and promoting personality growth.

R

receptor cell A specialized sensory cell, designed to pick up and transmit sensory information.

receptor molecule A specific protein on the surface of or inside a cell with a characteristic chemical and physical structure. Many neurotransmitters and hormones exert their effects by binding to receptors on cells.

reuptake A process by which released neurotransmitters are absorbed for later reuse.

rod A sensory neuron located in the periphery of the retina. The rod is sensitive to light of low intensity and is specialized for night-time vision.

S

schizophrenia A chronic mental disorder characterized by psychosis (e.g. hallucinations and delusions), flattened emotions, and impaired cognitive function.

second messengers Substances that trigger communications among different parts of a neuron. These chemicals play a role in the manufacture and release of neurotransmitters, intracellular movements, carbohydrate metabolism, and processes of growth and development. The messengers' direct effects on the genetic material of cells may lead to long-term alterations of behaviour, such as memory and drug addiction.

sensory neuron Neurons that send information from the sensory receptors (such as skin, ears, nose, and mouth) toward the central nervous system. They have two processes extending from the soma.

serotonin A monoamine neurotransmitter believed to play many roles, including, but not limited to, temperature regulation, sensory perception, and the onset of sleep. Neurons using serotonin as a transmitter are found in the brain and gut. Several antidepressant drugs are targeted to brain serotonin systems.

short-term memory A phase of memory in which a limited amount of information may be held for several seconds or minutes.

soma The body of a cell.

stem cells Cells with the potential to develop into many different cell types in the body. Serving as a sort of repair system for the body, they can theoretically divide without limit to replenish other cells as long as the person or animal is still alive. When a stem cell divides, each new cell has the potential to either remain a stem cell or become another type of cell with a more specialized function, such as a muscle cell, a red blood cell, or a brain cell.

stimulus An environmental event capable of being detected by sensory receptors.

stroke An impeded blood supply to the brain. Stroke can be caused by a rupture of a blood vessel wall, an obstruction of blood flow caused by a clot or other material, or pressure on a blood vessel (as by a tumour). Deprived of oxygen, which is carried by blood, nerve cells in the affected area cannot function and therefore die. Thus, the part of the body controlled by those cells cannot function either. Stroke can result in loss of consciousness and death.

strong artificial intelligence (or strong AI) Asserts that the brain is a kind of computer and the mind a kind of computer program. Strong AI aims to build machines that think. (The term "strong AI" was introduced for this category of research in 1980 by the philosopher John Searle of the University of California at Berkeley.) The ultimate ambition of strong AI is to produce a machine whose overall intellectual ability is indistinguishable from that of a human being.

subthalamus Represented mainly by the subthalamic nucleus, a lens-shaped structure lying behind and to the sides of the hypothalamus.

sympathetic nervous system A branch of the autonomic nervous system responsible for mobilizing the body's energy and resources during times of stress and arousal.

synapse A physical gap between two neurons that functions as the site of information transfer from one neuron to another.

T

temporal lobe One of the four major subdivisions of each hemisphere of the cerebral cortex. The temporal lobe functions in auditory perception, speech, and complex visual perceptions.

thalamus A structure consisting of two egg-shaped masses of nerve tissue, each about the size of a walnut, deep within the brain. The key relay station for sensory information flowing into the brain, the thalamus filters out information of particular importance from the mass of signals entering the brain.

V

ventricles Comparatively large spaces filled with cerebrospinal fluid. Of the four ventricles, three are located in the forebrain and one in the brainstem. The lateral ventricles, the two largest, are symmetrically placed above the brainstem, one in each hemisphere.

W

Wernicke's area A brain region responsible for the comprehension of language and the production of meaningful speech.

INDEX

Note: Where more than one page number is listed against a heading, page numbers in bold indicate significant treatment of a subject. Page numbers in italic refer to illustrations.

Britannica®

Since its birth in the Scottish Enlightenment Britannica's commitment to educated, reasoned, current, humane, and popular scholarship has never wavered. In 2008, Britannica celebrated its 240th anniversary.

Throughout its history, owners and users of *Encyclopædia Britannica* have drawn upon it for knowledge, understanding, answers, and inspiration. In the Internet age, Britannica, the first online encyclopædia, continues to deliver that fundamental requirement of reference, information, and educational publishing – confidence in the material we read in it.

Readers of Britannica Guides are invited to take a FREE trial of Britannica's huge online database. Visit

https://brain.britannicaguides.com

to find out more about this title and others in the series.